Exploring Solar Energy II:

Activities in Solar Electricity

Allan Kaufman
Old Dominion University
Norfolk, Virginia

◖▷ **Prakken Publications, Inc.**

*This book is dedicated
to my mother,
Jean Kaufman*

Acknowledgments

Many of our endeavors, especially those involving such things as books, would never get done without the encouragement, advice, and friendship of those around us.

I would like to extend personal thanks to the following individuals:

Mrs. Emily Jones, secretary of Old Dominion University's Occupational and Technical Studies Department, for her expert typing of the manuscript.

Susanne Peckham, book editor, Prakken Publications, for her expertise and guidance.

Jenifer Huismann, sales and marketing, Siemens Corporation, for photographs and technical documents.

Richard Blieden, president, B & E Energy System, Inc., for technical materials.

Dr. Robert Adams, vice-president, Solar Car Corporation, for photographs and technical documents.

R. Alan Panton, solar product manager, Kyocera Co., for technical materials and the cover photograph.

Dorothy Bergin, marketing and sales, Mobil Solar Energy Corporation, for photographs of current solar cell manufacturing equipment.

Contents

Acknowledgments   v
Preface   ix

Chapter 1   1
Photovoltaics: Applications and Systems
Chapter 2   22
Theory: Sunlight into Electricity
Chapter 3   37
Materials and Processes
Chapter 4   47
Solar Cells For the Experimenter

Project Section
Introduction   51
Chapter 5—Solar Sun Racer Vehicle   52
Chapter 6—Solar-Powered Rover Robot   59
Chapter 7—Solar Music Box   67
Chapter 8—Solar Rechargeable Flashlight   72
Chapter 9—Solar-Powered FM Transmitter   79

Appendix A   86
Efficiency of Solar Cells
Appendix B   89
Types of Photovoltaic Systems
Appendix C   90
Selected Bibliography
Appendix D   92
Solar Energy Terminology
Appendix E   96
Directory of Suppliers

Index   99

Preface

Electricity is a uniquely valuable, controllable form of energy that is fundamental to quality of life in the modern world. Miniature communication devices, laptop computers, and advanced medical instrumentation are but a few of the technologies that have improved our lives. All depend on this marvelous source of power, electricity.

Photovoltaics, the direct conversion of light into electricity, is an elegant and reliable source of energy for these new technologies and for the expanding world population. Photovoltaic cells that make use of solar energy can work in almost any geographic location and operate in a variety of harsh environmental conditions. These cells have no moving parts to wear out or break down, and they will produce electricity without noise, combustion, or pollution. Most important, their operation requires no coal, oil, or gasoline. The "fuel" involved—sunlight—is free!

As in my first book, *Exploring Solar Energy: Principles and Projects*, this text provides teachers, students, and interested consumers with a fundamental understanding of solar electricity. The book explores photovoltaics from an easy-to-follow conceptual format, and it provides several interesting solar electric projects.

Consumers and representatives of industry and government face a formidable challenge in formulating a coherent policy that integrates technical innovations with economic considerations in ways that can capture the potential of solar electricity. Many are rising to meet this challenge—which is encouraging for all of us.

—*Dr. Allan Kaufman*
Associate Professor
Old Dominion University

CHAPTER ONE

Photovoltaics: Applications and Systems

THE photovoltaic (PV) cell is an ingenious device that converts light into electricity (Figure 1-1). Unlike many nonrenewable sources of energy, solar cells are reliable, nonpolluting, and energy-efficient. They operate from a seemingly infinite supply of energy—the sun[1].

The direct conversion of light into electricity, the *photovoltaic effect*, was first reported by the French scientist, Alexandre Edmond Becquerel in 1839. Becquerel placed two metal electrodes into an electricity-conducting solution, called an electrolyte, and observed that the current that was produced varied with the intensity of the light. This observation was the beginning of many significant historical developments in the field of solar electricity that have taken place between 1839 and the present.

From 1873 to 1905, research significantly advanced humanity's knowledge of photoconductive materials. Scientists discovered that selenium was a light-conductive material, and that in a solid form it could produce the photovoltaic effect. Substances such as copper-cuprous-oxide were also found to be light sensitive.

Around 1905, Albert Einstein was advancing his theory of relativity. A portion of his theory had significant implications for the field of solar electricity, since it addressed the nature of light and light's effect on materials. Einstein theorized that photons, particles of light, could drive electrons out of materials, and that the

Figure 1-1.
Photovoltaic cells
convert sunlight
directly into
electricity. They are
manufactured in a
variety of shapes and
have a typical output
of .5 V at 300 mA.

Photo courtesy of Siemens Solar Industries

energy of the electrons varied with the light's wavelength. We know today that photons do interact with electrons of light-sensitive materials to create electron-hole movement, more commonly referred to as electricity.

The 1940s and 1950s were marked by advances in the development of several new photovoltaic materials and products. Companies such as Bell Laboratories and Western Electric explored light-sensitive materials such as single-crystal silicon, gallium-arsenide, and indium phosphide. Businesses tried to develop and market a variety of solar-powered products, such as radios, toys, and highway construction warning flashers. However, most of these devices could operate less expensively on conventional power sources. Despite the economic problems of solar electric power, two successful products did emerge that were powered by photovoltaics—a dollar bill changer and a machine that decoded punch cards and tape.[2]

Although more than 100 years had passed since the photovoltaic effect had been discovered, economically viable applications were not possible until the advent of the United States' space program in the 1960s. Common energy sources—batteries, oil, and gas—that were plentiful and inexpensive to use on earth proved too expensive or impractical for use in space. The photovoltaic cell fulfilled the need for a reliable, lightweight source of energy for America's satellites (Figure 1-2). Throughout the 1960s, PV cells demonstrated their ability to function in spite of the great temperature variations and high radiation levels found in the harsh environment of space. The space program helped to move solar cells out of the scientist's laboratory and into practical application.

The photovoltaic cells used in the space program were basically made by hand and cost approximately $1,000 per peak watt of energy—far too great a cost for widespread use in residential and commercial applications. Throughout the 1970s, 1980s, and into the 1990s considerable emphasis has gone toward cost reduction. Companies such as Siemens Solar Industries, Solarex Corporation, Applied Solar Energy, and Mobile Solar are all attempting to create a cost-effective photovoltaic system.

Efforts to make a PV cell cost competitive with conventionally generated electricity have progressed. In the 1950s, the cost per watt of peak generating capacity (the power a cell produces at noon on a sunny day) was about $2,000. In the 1970s, the cost dropped to approximately $100 to $200 per watt, and today the generating cost is about $5 per watt, or about 30¢ per kilowatt-hour.[3] This is still considerably more than the 10¢ per kilowatt-hour from the nearby utility connection, but it is estimated that by

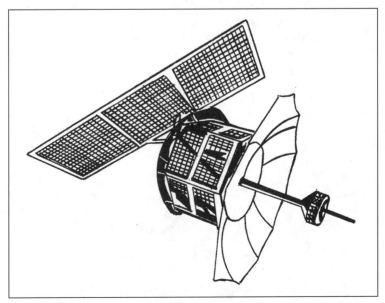

Figure 1-2. The primary use of photovoltaic cells in the 1960s was to provide power for the many new satellites in the emerging U.S. space program.

the year 2000, PV electricity generated for large-scale utility applications will be no more than 12¢ per kilowatt-hour![4]

Solar electricity will become a more attractive energy source as it becomes more economical. The research and development efforts of private industry are being guided by this goal. In 1990, Dr. Charles Gay, president of Siemens Solar Industries commented on this economic goal, "Economy is one of the main ideas driving this industry. . . . I suspect that in the course of the next decade, solar energy will become competitive with other sources of bulk power. In my opinion, it is inevitable that solar energy will become a significant contribution to the mix of electric power and supply throughout the world."[5]

There are various present day, cost-effective uses for solar electricity. However, since photovoltaic cells do not generate electricity unless light strikes them, it is important to briefly discuss the source of solar energy, the sun.

Solar Electricity— The Source

Formed from a cloud of gas primarily composed of hydrogen atoms, the sun constantly converts its mass into energy by a fusion process. The sun has been in this hydrogen-burning process for about six billion years, and it will continue to radiate energy for billions of years to come (Figure 1-3).

Energy leaves the sun and reaches the earth in the form of *wavelengths*. (See Chapter 2 for a discussion of wavelengths and sun-

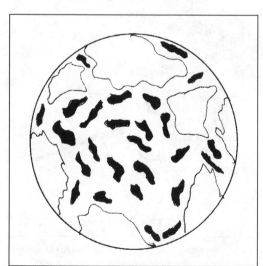

Figure 1-3. The sun, the source of solar energy, will continue to radiate energy for billions of years. Its core temperature is 18,000° F and its surface temperature is 10,000° F.

light.) The amount of the sun's energy that reaches the edge of the earth's atmosphere is called the *solar constant*, and it is equal to 1.353 kW/m² (Figure 1-4). Since the atmosphere blocks and diffuses some of this energy, the solar constant is slightly less at sea level, about 1.000 kW/m². A square meter is a small area, yet if we

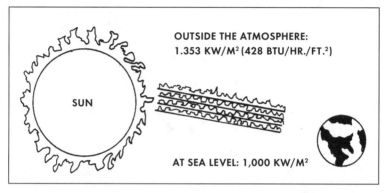

Figure 1-4. The solar constant. If we placed one square foot of glass outside the earth's atmosphere, it would intercept 428 BTUs each hour. This would equal 1.353 kW of electricity per square meter of area. Scientists refer to this energy as the solar constant.

could convert into electricity all of the sunlight that strikes a square meter area in a 1-hour period of time, we could light a 100 W bulb for 10 hours, or smelt enough aluminum to produce a six-pack of soda cans or heat enough water for a shower.[6]

Unfortunately, not all of the radiation leaving the sun can be converted into electricity. Photovoltaic cells currently are capable of conversion efficiencies of only approximately 14 to 16 percent (see Chapter 2 for a detailed discussion), and the earth's atmosphere reflects about 35 to 40 percent of the sun's radiation (Figure 1-5).

Some solar radiation is reflected by the clouds and, thus, never reaches the photovoltaic cells. Another portion, 15 percent, is absorbed into the ozone layer, water vapor, and carbon dioxide of the atmosphere. Fortunately, enough direct and diffuse solar radiation still strikes a solar cell array during a typical day to create a substantial amount of electricity.

We have all experienced an interesting situation that determines the amount of solar radiation that will strike a solar cell array. In the early morning hours, the sun's radiation feels less intense because the sun is close to the horizon and the path of its radiation

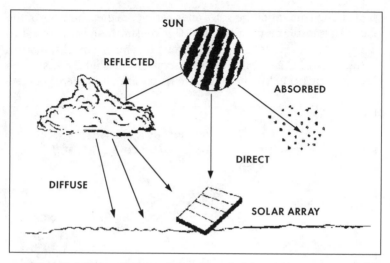

Figure 1-5. Forms of solar radiation. Not all of the sun's energy is available for conversion to electricity. The earth's atmosphere reflects about 35 to 40 percent of the sun's radiation.

through the earth's atmosphere lengthens. At noon, with the sun directly overhead, solar radiation travels through less of the atmosphere, and, thus, we receive more direct radiation. During sunset hours, the path once again lengthens as the sun approaches the horizon. The more atmosphere the sun's radiation must pass through, the *lower* the energy content we receive.

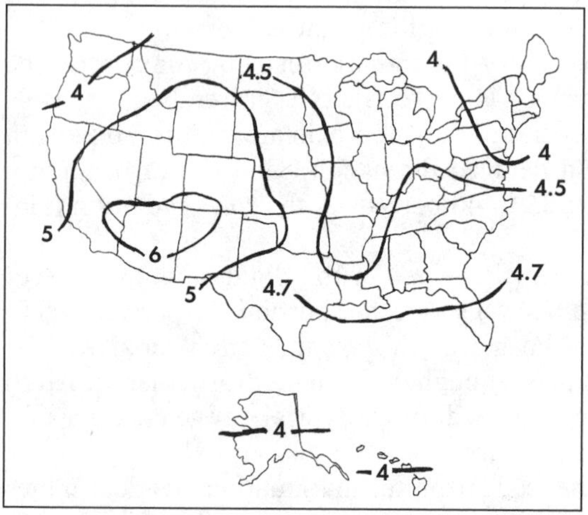

The earth's rotation about its axis and its revolution around the sun also affect the amount of solar energy available to photovoltaic cells. The earth's rotation produces hourly variations in the power intensity of the sun at dif-

Figure 1-6. Engineers rely on sun maps that illustrate the average peak sun hours each day (or year) for a particular geographic region of the country.

ferent locations on the ground. The desert area of the southwestern United States consistently receives a great deal of sunshine and is an excellent geographic location for solar electric arrays.

Fortunately, solar engineers can use established meteorological data to aid them in planning the use of solar electric arrays. The National Climatic Data Center located in Ashville, North Carolina, provides average hourly insolation (sun intensity) data through a statistical analysis of 26 cities. Other important sources of information are *peak sun hour* maps (Figure 1-6). Peak, or full, sun is approximately 1,000 W/m² on a clear day. Using special maps, such as the one shown in Figure 6, engineers can determine the number of peak sun hours per day, or average per year, that a particular geographic area will receive. For example, on a typical day in Virginia, the maps indicate about 4.5 average peak sun hours each day.

Sunlight and geography are only two elements that must be considered in the design of a solar electric system. If photovoltaic electricity is to become a successful partner of or replacement for conventional sources of energy, factors such as the technical characteristics of solar cells and the economic feasibility of their applications must be examined.[7]

Solar Electricity—A Spectrum of Applications

At present, there exists an interesting variety of cost-effective

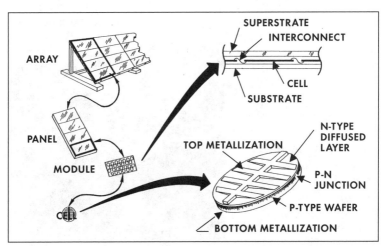

Figure 1-7—Photovoltaic array nomenclature. Solar cells are electrically connected to modules, which are then assembled and connected into panels. The panels are grouped together to form a solar array. Arrays range in size from two or three panels to large systems using hundreds of panels.

applications of photovoltaic systems. There are three broad market segments of primary importance:

- stand-alone systems,
- the remote market (the developing world), and
- industrial/residential systems.

The stand-alone photovoltaic system is designed for applications

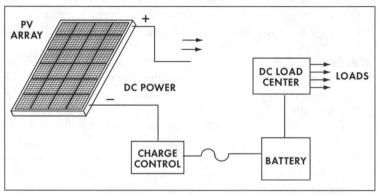

Figure 1-8. The stand-alone photovoltaic system is one of the simplest solar designs. It is particularly useful in remote regions that lack easy access to utility lines.

that must be self-sufficient. The essential components of this system are (1) a photovoltaic array (Figure 1-7), (2) a rechargeable storage battery, (3) a charge regulator, and (4) a suitable direct current load (Figure 1-8).[8] All of the components are matched in such a way that the electricity produced from the sun will meet reliably the application's demand for energy 24 hours per day, every day of the year.

Since the stand-alone system (Figure 1-9) operates reliably in all types of weather and requires very little maintenance, it has found extensive use in

Figure 1-9. A small mobile photovoltaic array comprised of 10 solar electric panels. Such a system is one component of the stand-alone solar electric design.

remote regions where fuel delivery would prove difficult or utility power lines do not exist. Standalone systems are frequently used to power radio and television transmitters since their ideal sites are located at very high elevations where utility lines are not available (Figure 1-10). On the island of Curacao in the Caribbean, solar electricity provides 100 percent of the power for Radio Hoyer's FM transmission (Figure 1-11). Solar electricity is often economically competitive with conventional energy sources that are located in island communities.

Providing electricity for operations at sea has been difficult in the past. Offshore drilling rigs have relied on batteries and generators that require maintenance, fuel, and replacement. Stand-

Photos courtesy of Siemens Industries.

Figure 1-10. A South American solar-powered transmitter. Photo voltaics is ideal in remote, high-elevation regions.

Figure 1-11. A solar-electric-powered FM radio transmitter located in Curacao.

Figure 1-12. The U.S. Coast Guard uses solar electricity to power its navigational aids.

alone photovoltaics are capable of powering appliances and transmission equipment on these marine systems. Also, the United States Coast Guard uses small solar electric panels to power buoys that it scatters along U.S. waterways (Figure 1-12).

Once considered expensive, exotic, scientific toys, photovoltaic cells and modules are now economically used in consumer and leisure products. Hand-held solar-powered calculators once cost about $50; today, these de-

Figure 1-13. Photovoltaics can be applied to consumer products such as watches and calculators. Small devices use miniature solar cell panels called *microgenerators*.

vices sell for less than $20. The calculators (Figure 1-13) are powered by a series of small solar cells connected together to create

Figure 1-14. Photovolatic kits are available to the consumer to mount on recreational vehicles (RVs). The solar cells will power an assortment of appliances.

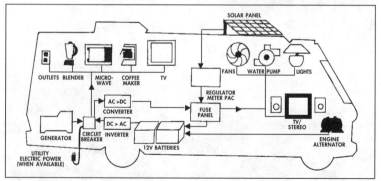

Photo courtesy of Siemens Solar Industries

Figure 1-15. Schematic drawing of a solar-electric-powered recreational vehicle. All of the appliances, from lights to stereo system, are powered by the sun. Similar solar electric systems are used on sailboats and houseboats.

several volts of electricity. These devices will operate in sunlight, as well as under incandescent or fluorescent light.

The photovoltaic cells used to power electronic products such as watches or calculators are very small (about 3/8" x 1/2" for a calculator). These units are often referred to as microgenerators and are frequently used to recharge small batteries for products or to provide direct electrical energy. Solar microgenerators efficiently produce electricity at both high and low light levels and they are easy to integrate into the appearance of the product.

Photovoltaic systems used in consumer/leisure products vary in size, power, and application. An interesting use comes from their ability to make recreational vehicles electrically independent. As illustrated in Figure 1-14, a photovoltaic array can be mounted on the top of a motor home. During the day, the solar cells produce enough electricity to recharge the recreational vehicle's (RV) 12 volt batteries, eliminating many hours of generator use.[9]

Figure 1-15 shows a solar electric recreational vehicle. The solar panels mounted on top of the RV will supply 15 to 16.9 volts at 34 to 51 watts of energy.[10] The high-powered photovoltaic chargers generate enough electricity to run everything from lights, TVs,

pumps, stereos, and furnace blowers. A similar system is used to extend battery life and provide extra power on sailboats and houseboats.

There are numerous practical and economical applications

Figure 1-16. Photovoltaic modules, shown above, charge the batteries that supply energy to light billboards at night.

Figure 1-17. Located far from utility power lines, the billboard shown at right is powered by solar cells.

Photos courtesy of Siemens Solar Industries

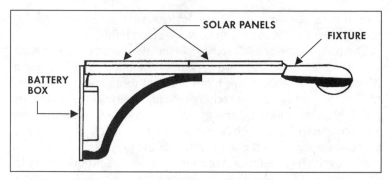

Figure 1-18. The Solarpal™ safety streetlight is an energy-efficient, solar-electric-powered light. It is useful for residential area lighting and will operate five consecutive days without the sun.

of photovoltaics in the "commercial" product area. Solar-electric-powered telephone systems, bus shelters, marinas, and post lighting are just a few of the diversified applications. As illustrated in Figures 1-16 and 1-17, solar cells are used to recharge batteries that provide nighttime lighting for billboards located far from utility lines.

An interesting and useful solar-electric commercial product is the Solarpal™ safety streetlight (Figure 1-18).[11] This highly reliable solar-powered lighting system is a cost-competitive alternative for the area illumination of residential streets, parking lots, and general security lighting. The device is a self-contained system that mounts on a pole or wall. During the day, the solar panels charge a gel-type battery that is housed in a sealed box. As evening approaches, reception of the sun's energy decreases to a point where the electronic circuit turns on the light. No utility connection is required, no maintenance is needed, and the unit will operate five consecutive days without receiving sunlight.

One of the promising applications for use of photovoltaics is in remote villages throughout the United States, as well as in developing countries. The Indian Health Service in the U.S. estimates that about 80,000 individual Native Americans' homes do not have electricity. Most of these homes are in villages that are far from the local utility power grid. In 1978, the U.S. government developed an experimental, solar-electric program to aid people in these villages.

For example, the Schuchuli Indian village, located in western Arizona, had no electricity prior to 1978. The village was 17 miles from the nearest utility lines. Its 15 families had to rely on candles for lighting and a diesel-powered pump for water. A small 3.5 kW photovoltaic system installed in December 1978 powered lights and refrigerators in the village's homes and lights in the community feast house, as well as a washing machine and a sewing machine.

The United Nations estimates that there are about 10 million villages like Schuchuli throughout the world—in other words, vil-

Figure 1-19. Water-pumping system powered by a photovoltaic array. Solar electricity aids in irrigating in remote regions.

Photo courtesy of Siemens Solar Industries

lages that lack electricity for even the basic necessities of life.[12] Solar electricity has great potential for improving the quality of human life worldwide.

Today, photovoltaics can make a cost-effective contribution to developed and developing countries, particularly regarding provision of water for basic needs. Countries in the Middle East, such as Israel and Saudi Arabia rely on solar-electric-powered systems for their desalination units, and photovoltaic-powered pumps bring water to agricultural areas in those countries (Figure 1-19). Whether through a small-size photovoltaic system providing the electricity for a water-pumping site in Papua, New Guinea (Figure 1-20), or a large utility-scale system for entire communities (Figure 1-21), solar electricity can play the role of an economical partner to conventional energy sources.

Residential Photovoltaics

Building a home with a self-sufficient energy source has long been a dream of architects and engineers. During the past 10 years, the cost of photovoltaic systems has gone down, giving architects an opportunity to construct small, stand-alone 50 to 500 watt so-

Photos courtesy of Siemens Solar Industries

Figure 1-20. Solar-powered water-pumping station in Papua, New Guinea.

lar-powered homes. The electrical systems in these small solar electric homes are often referred to as *cabin systems*. They are used in locations where the energy demand is so small that the cost of the required utility transformers and power lines is greater than the cost of using photovoltaic systems. Solar electric cabin systems have a residential market in northern California, Arkansas, New Hampshire, and Maine, where remote houses are not served by local utility companies.

If solar electricity is to become a significant force in the housing sector, the industry will have to build and sustain a robust PV residential market. The systems used to power a conventional home will have to be 5 to 10 times the size of the small stand-alone cabin units. A large-scale PV system must be capable of generating 50 to

Figure 1-21. Siemens produces solar electric modules for the largest utility-scale systems. They show solar electricity to be a reliable alternative to renewable energy sources.

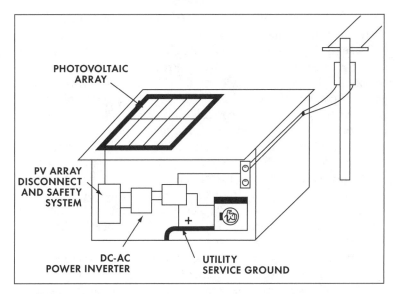

Figure 1-22. Schematic of a utility-interactive, photovoltaic system used to supply electricity for a typical home.

100 percent of the electricity used annually by a family that relies on a full complement of household appliances. The system must also operate compatibly with the local utility.

What type of components would be needed to create such a residential photovoltaic system? It is interesting to note that only two major items are required:

● a photovoltaic array which generates direct current (dc) and

● a power inverter (also known as a power conditioner), which converts the PV module's dc electricity to the house's alternating current (ac) electrical needs.

Some PV systems also use an on-site electric energy storage arrangement that uses lead acid batteries. Figure 1-22 illustrates a basic residential PV system.

Since a typical family's electrical energy needs will vary throughout the day there is a mismatch between the PV's electrical supply and the house's needs. A solution to this problem calls for the use of a utility-*interactive* photovoltaic system. When the home's electrical demands are greater than the output of the PV array, power is drawn from the local utility lines. When the house's load is less than the PV array's output, the excess electricity is run back to the utility system. The Public Utilities Regulatory Policies Act of 1978 requires utility companies to accept electricity from photovoltaic homes and to pay the home owner for the energy at a rate based on the marginal utility energy costs.

To send the photovoltaic house's electricity back to the utility grid, it is necessary to change the PV array's dc electricity into compatible ac power (120/240 V and 60 HZ cycle). The inverter creates this compatible ac power and serves as the electronic intelligence of the entire power-conditioning system. The inverter and its associated electronic circuitry monitor the power of the photovoltaic array to insure the availability of sufficient electricity for the residence throughout the day.

The power-conditioning system also serves as a safety control, monitoring the electricity in case of a power outage or an unusual problem in the utility system. The inverter's system will disconnect the photovoltaic array from the utility grid to protect utility workers from additional danger.

In 1982, the Solar Design Associates firm designed a photovoltaic, utility-interactive home in Massachusetts. The Carlisle Home (Figure 1-23) was one of several prototype solar electric projects

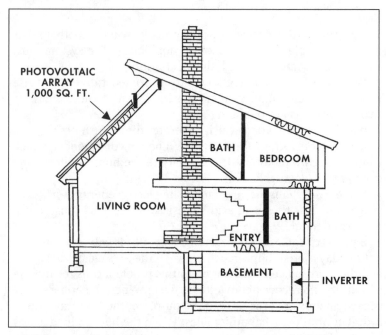

Figure 1-23. The Carlisle Home uses 126 photovoltaic modules measuring 2' x 4' to power household appliances. The residence generates excess electricity that it sells back to the local utility company.

constructed in the United States to develop photovoltaics for practical home use. The 3,100 square foot home required roof installation of 126 2' x 4' photovoltaic modules. Each module contained 72 silicon solar cells rated at 58 W of peak power. The total array could produce a daily output of 7.3 kW of energy.[13]

The Carlisle Home uses the utility-interactive system based on a dc to ac power conditioner. Inside the home is a full complement of appliances powered by the sun: heat pump, dishwasher, clothes washer and dryer, range and oven, and whirlpool bath. Even with all of these energy-consuming products, the house sells excess electricity back to the utility, the Boston Edison Company.

In 1982, when the Carlisle Home was built, photovoltaic systems cost $7.50 per watt. The home's solar electric array cost $75,000. Today, the cost of photovoltaics has dropped to $5 per watt, thus significantly increasing the feasibility of building a solar-powered home. When photovoltaics are reduced in cost to $1.60 per watt, a solar electric system similar to the one used on the Carlisle Home will cost about $10,000.

The Solar Electric Car

Production of an automobile powered by sunlight is an intriguing prospect. So far through the 1990s, significant emphasis by local, state, and the federal government has resulted in legislation (and pending legislation) containing provisions for alternative-fueled and electric vehicles.

The National Energy Security Act of 1991 and the Electric Vehicle Amendment adopted by the Senate Energy Committee in June 1992 authorized a research and development program on electric hybrid vehicles.[14]

At the state level, California has taken the lead role in developing certification procedures for electric vehicles. By 1998, at least 2 percent of all cars sold in California must be zero-emission vehicles, with a rise to 5 percent in 2001 and to 10 percent in 2003. In 1992, 15 states were considering the California program, and the governors of Massachusetts, New York, New Jersey, Maine, Maryland, and Pennsylvania have announced their support for California's regulations. Many of the quarter million zero-emission vehicles in the year 2003 may be solar electric powered.

The idea of using electricity to power a car is not new. The first electric car was built in 1838, and there were more electric cars on the road than gasoline-powered cars until the early 1900s. These early electric cars were battery powered; the concept of using pho-

Figure 1-24.
A solar-
electric-
powered
Ford Festiva

Photo courtesy
of Solar Car
Corp.

tovoltaic cells to convert sunlight into electricity to power a vehicle is new.

The solar electric car, once the province of the experimenter, is
now the basis of a new industry. The combination of environmental concerns, regulatory incentives, and technological breakthroughs
is creating an industry on the verge of mass-market status.

The majority of solar electric cars to date are based on a "retrofit" approach (Figure 1-24). The car illustrated in Figure 1-24 is a
Ford Festiva retrofitted into a solar electric vehicle by Solar Car
Corporation of Melbourne, Florida. The car's basic safety features
have been retained, while solar cells have been attached to the
hood and roof. The vehicle's top speed is 60 mph, and it has a

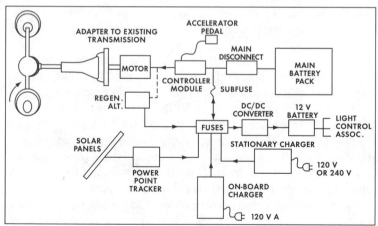

Diagram courtesy of Solar Car Corp.

Figure 1-25. A schematic diagram of a solar-electric-powered vehicle. Most
designs use solar cell modules in connection with batteries.

range of about 60 miles before its batteries need recharging. On a typical sunny day, the solar panels generate electricity to supplement the vehicle's range by 10 miles. The photovoltaic panels on the car are constantly trickle charging the batteries during daylight hours. (A basic solar electric system for a standard vehicle appears in Figure 1-25.)[15]

The solar electric Ford Festiva looks like a typical two-door hatchback, except that there are solar cells on the roof and hood and the car has no carburetor, spark plugs, or exhaust system. Although the cost of the solar electric Festiva is twice that of a conventional model, the car would save about $1,000 per year in oil and gasoline. To operate the car, the cost is only two cents per mile, compared to seven or eight cents per mile for a gasoline-powered vehicle. If the car was mass-produced, its price could compete with the price of gasoline-powered vehicles.

To make a solar electric car that performs like a gasoline-powered car, a fundamental challenge must be addressed: creating an effective battery for the system. The standard lead acid battery now used in gasoline autos puts out too little energy per pound to move the car unless a great deal of the car's space is filled with batteries.

In May 1992, Ford, General Motors, Chrysler, and the federal government created the Advanced Battery Consortium. The consortium awarded a contract to the Ovonic Battery Company of Troy, Michigan, which is developing a nickel-metal hybrid battery. The

Photo courtesy of Kyocera Company

Figure 1-26. Solar-electric-powered sun racer vehicles are designed and raced by engineering students from 32 universities in the Sunrayce U.S.A.

company claims this battery will propel a small car 300 miles on one charge. The small car will go 0 to 60 mph in 8 seconds, reach a top speed of 100 MPH, last 100,000 miles, and recharge in 15 minutes.[16]

If solar electric cars are to become commonplace in our society, the consumer must be informed of recent technological progress made with batteries and solar cells. Public awareness of solar electric vehicles is being promoted by General Motors and the Society of Automotive Engineers. The GM Sunrayce is an outgrowth of the World Solar Challenge, a triennial competition held in Australia. Sunrayce USA gives student engineers from 32 universities and colleges throughout the U.S., Canada, and Puerto Rico the opportunity to demonstrate the potential of solar energy (Figure 1-26). The students design, test, build, and race prototype solar electric cars.

Solar-powered vehicles will become a reality in the near future. The U.S. Postal Service and Federal Express Co. are seriously looking into electric vehicles, and some corporate fleets already have some vehicles that have been converted to this new technology.

Notes

[1] The sun does have a defined lifetime; however, it loses such a small percentage of its mass each second that it is expected to continue to radiate energy for billions of years.

[2] Paul Maycock and Edward Stirewait, *Photovoltaics: Sunlight to Electricity in One Step* (Andover, MA: Brick House, 1981).

[3] At the current price of 30¢ per kWH, it is less expensive to install solar cells on a home that is one-half mile from the utility grid than it is to extend and connect power lines.

[4] Edward Edelson, "Solar Cell Update," *Popular Science* (June 1992): 95.

[5] Dietrich Stahl, "Products with a Place in the Sun," *Siemens Review* (Siemens Solar Industries) 57, No. 6 (November/December 1990): 7.

[6] Arnold Fickett, Clark Gillings, and Amory Lovins, "Efficient Use of Electricity," *Scientific American* (September, 1990): 65.

[7] See Chapter 2 for a discussion of the technical characteristics of photovoltaic cells.

[8] Photovoltaic cells produce direct current electricity because electrons flow in one direction only.

[9] A solar panel produced by Kyocera Company will replace four hours of generator use.

[10] This system is designed by the Kyocera Company, located in California.

[11] Manufactured by the Solar Outdoor Lighting Company.

[12] See note 2 above.

[13] Richard Stepler, "Solar Electric Home I," *Popular Science*, September 1981.

[14] Seven bills were introduced into the 102nd Congress that dealt with alternative-powered vehicles.

[15]Solar electric cars are powered by batteries that are recharged by photovoltaic panels and a standard 110 V socket.

[16]Boyce Rensberger, "Consortium Seeks New Battery for Autos of the Future," *The Washington Post*, 19 June 1992.

CHAPTER TWO

Theory:
Sunlight into Electricity

To understand how the sun's energy is converted into electricity, we must examine the fundamental nature of sunlight. For many years, scientists involved in the study of particle physics suggested that sunlight basically comprised *wavelengths* of energy. However, some scientists, notably Albert Einstein, argued that light was really composed of small, individual spheres of energy called *photons*.

We know today that the wavelength and photon theories are both useful in describing sunlight (Figure 2-1). Photons are corpuscles of energy that possess mass, travel at a speed of 186,000 miles per second, and stream from the sun in various wavelengths. Some of these wavelengths make up our visible spectrum, colors such as red, blue, and yellow. Others—ultraviolet and infrared wavelengths—are invisible to the human eye (Figure 2-2).

Although these wavelengths may be interpreted by the layperson as simply "color," scientists have determined that wavelengths of photons exist at different energy levels. It is the energy level of the photon that plays an important role in converting sunlight into electricity.

All photons leave the sun traveling at the same rate of speed. However, since some of these corpuscles of energy have greater mass than others, their energy level is greater. Photons that possess enough energy will favorably interact with certain *light-sensitive* materials to produce electricity.[1]

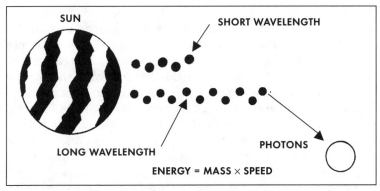

Figure 2-1. Light streams from the sun in the form of small corpuscles of energy called photons. Photons that possess sufficient energy will favorably interact with light-sensitive materials to produce the photovoltaic effect.

Light-sensitive materials have been used in electronic components, such as transistors, for many years. They are commonly referred to as *semiconductors* because they are neither good conduc-

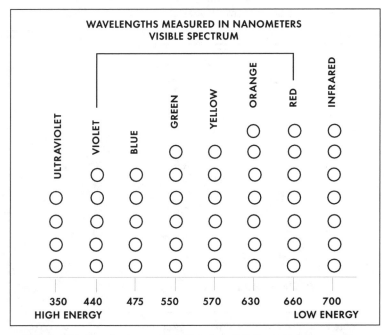

Figure 2-2. The shorter wavelengths of sunlight have high-energy photons while the larger wavelengths, such as infrared, have low-energy photons. Fortunately, solar cells respond to a broad range of light wavelengths.

tors nor good insulators of electricity.[2] Certain semiconductors, such
as silicon, are frequently used in the photovoltaic process.

To function in the photovoltaic process, silicon must be refined
into a pure crystalline state. In this state, silicon's atoms assume an
ordered cubic arrangement, but this structure does not permit
silicon's outer-shell electrons (particles of energy in atoms) to move
easily from atom to atom, which is a necessary condition for creating electricity. However, when the crystalline silicon atom is bom-
barded by a photon (sunlight) of sufficient energy, an electron is
forced out of its position, creating an empty "hole" in the lattice[3]
(Figure 2-3). The electron possesses a negative charge, and its hole
has a positive charge. The electron moves around the silicon's lattice searching for its positive hole. At the same time, other electrons energized by photons fill vacant holes, one after another.[4]
This process, commonly referred to as *electron-hole movement,* forms
the basis for an electric current.

Refining the Process

A fundamental problem exists in the pure silicon's atoms that
precludes its use as a generator of electricity. The negatively charged

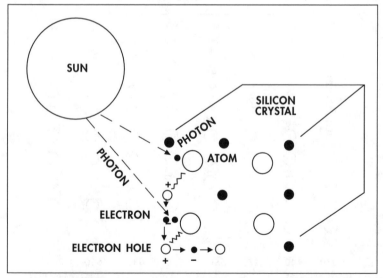

Figure 2-3. When a photon strikes a crystalline silicon atom, it frees an outer-
shell electron (negative charge) and creates an empty "hole" (positive charge)
in the lattice. The process occurs throughout the crystal of silicon and is known
as *electron-hole movement,* the basis for electricity.

electron and its positively charged hole do not stay apart very long. Unless some provision is made in the pure silicon crystal, the electrons and their original holes will reunite in about a millionth of a second. This process is referred to as the *recombination* of the electron and its hole.

Fortunately, scientists have developed a method to keep the negative and positive charges apart and to capture their energy as electricity. The process involves "doping" the pure silicon with a tiny amount of two impurities: boron and phosphorous. It uses about 1 part boron per 100,000 parts silicon, and the phosphorous only penetrates to a depth of 300 to 600 atoms.[5]

The addition of these two elements upsets silicon's ordered lattice (Figure 2-4). Because a boron atom possesses only three outer-shell electrons, it lacks one electron for each bond in the lattice. This part of the silicon wants to accept electrons to fill its positive holes. Consequently it is called *P-type* silicon.

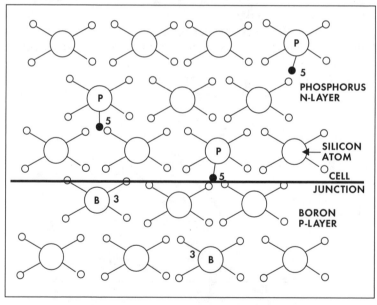

Figure 2-4. Doped silicon crystal. Small amounts of boron and phosphorus are added to the pure silicon. Boron lacks an extra electron, while phosphorus has one more electron than it needs for each bond in its structure. The cell junction allows only the high-energy electrons in the phosphorus layer to pass through it to the boron side when the silicon cell is exposed to sunlight. Since electrons can flow only from the positive P-layer (phosphorus) to the negative N-layer (boron), a difference of potential is created in the cell, which is called voltage. Diagram courtesy of Mobile Solar Energy Corp., Billerica, MA.

The phosphorus-doped portion of the silicon crystal possesses atoms with five outer-shell electrons. Each phosphorus atom has an extra electron, one more than it needs to satisfy each bond in the structure. Since this part of the silicon has an excess of free electrons, it is *negative* in nature and is called *N-type* silicon.

The doped silicon crystal now has a positive and negative side (similar to a battery's two poles). The two sides of the cell meet at a "junction" that is created during the time of manufacture (the junction is only 0.00005 of an inch thick).[6] This junction is really a region of a static electrical charge, similar to static electricity, and it serves as an energy barrier in the silicon solar cell.

The cell junction only allows the "high" energy electrons in the positive P-silicon side to pass through it. The electrons in the negative N-silicon are "low" energy and are blocked by the static electrical charge. Thus, the electrons may flow only from the positive P-layer side of the cell to the negative N-layer side. This phenomenon creates a difference of potential between the two sides of the silicon solar cell called *voltage*.

If we place the solar cell in sunlight (Figure 2-5) the photons will cause the high energy electrons in the P-layer to travel across the cell's junction into the N-layer. The electrons now accumulate in the N-layer. If we connect a wire from this portion of the cell to a motor, then connect a wire from the motor to the P-layer, the

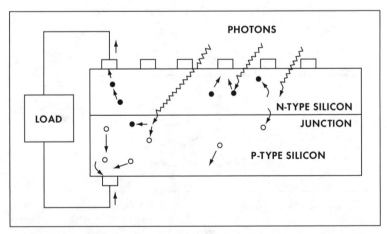

Figure 2-5. Solar cell in sunlight. Composed of photons, sunlight acts like a pump that constantly frees electrons in the P-layer to travel across the junction into the N-layer. Electrons will accumulate in the N-layer, and if a wire is connected to the cell a current will flow to the load. Diagram courtesy of Mobile Solar Corporation.

electrons will travel from the N-side of the cell through the motor and return to the P-side where they will fill vacant electron holes.

The sunlight (photons) acts as a pump, freeing electrons that eventually travel through the silicon cell and through the circuit as current that can power energy needs—thus becomming a light-activated battery, the photovoltaic cell.

Characteristics of the Photovoltaic Cell

The production of electricity from a photovoltaic cell, or silicon solar cell, is an elegant process (Figure 2-6). Streams of photons provide the initial push to free electrons in the cell, which then move through an attached wire as current powering a variety of our energy needs. Unlike a battery, which undergoes chemical changes as it produces electricity, a photovoltaic cell emits no chemicals or pollutants, consumes no materials in its energy conversion process, and can be manufactured with an effective lifetime of more than 20 years.[7]

The energy output of a solar cell varies with the amount of available light. Unfortunately, not all of the sunlight that strikes a photovoltaic cell can be converted to electricity. Scientists suggest that there is a *theoretical* limit to the quantity of energy we can get out of a solar cell in relation to the energy in the sunlight that strikes the cell. This theoretical limit, or energy conversion efficiency, is about 25 to 30 percent; however, the *practical* efficiency of a silicon solar cell is even lower, approximately 12 to 16 percent.[8]

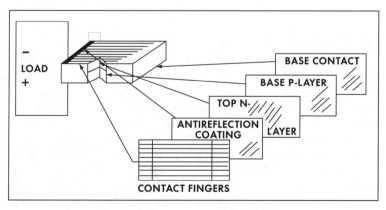

Figure 2-6. A typical silicon solar cell comprises a base contact, a positive P-layer, a negative N-layer, an antireflection coating, and a set of contact fingers. The contact fingers must be designed so they do not block sunlight from striking the cell.

What prevents a solar cell from converting all of the sun's energy into electricity? Several basic factors affect a cell's efficiency level. First, some materials are not as efficient as others in the conversion process. Polycrystalline, although less expensive than pure silicon, loses some efficiency in its tiny silicon grain boundaries. Second, the photovoltaic cell's surface tends to reflect sunlight. This problem is partially addressed by coating the cell's surface with a thin antireflection material, similar to the coating used on camera lenses. Finally, the electrical grid on the N-layer of the cell must be designed and manufactured so that it does not block the sunlight from reaching the active layers.

Material	Maximum Conversion Efficiency
Germanium	13 percent
Cadmium Sulfide	18 percent
Silicon	25 percent
Cadmium Telluride	25 percent
Indium Phosphide	26 percent
Gallium Arsenide	27 percent
Aluminum Antimonide	27 percent

Source: Paul Maycock and Edward Stirewalt, *Photovoltaic: Sunlight to Electricity in One Step,* Brick House Publ., Andover, MA, 1981, p. 37.

Although it may appear that photovoltaic cells are not very efficient energy-conversion devices, we must note that the automobile engine's energy conversion efficiency is only about 20 percent.

Electrical Characteristics

Since a photovoltaic (PV) cell does not convert all of the sunlight striking it into electricity, and since each day may be different with respect to the intensity of sunlight (irradiance) and temperature, engineers need to know how the electrical characteristics of solar cells are affected by these variables.

Using a series of I-V (current-voltage) curves on a graph, we can evaluate several important variables. If we place a silicon cell in bright sunlight, but do not connect it to a load, such as a motor or other appliance, an *open circuit will exist,* one that has infinite resistance. Since no current can flow in an open circuit, the PV cell's current value would be zero. However, if we measured the *voltage* between the positive and negative terminals of the cell, we would find about .570 volts.

What would happen to the current and voltage if we connect the PV cell to a product? Although the appliance will offer some resistance in the circuit, it will be less than the infinite resistance of the open circuit. As the resistance begins to decrease, the voltage in the circuit will begin to decrease to about .45 volts; however, the current will start to increase!

If we continue to decrease the appliance's resistance value, the solar cell's voltage also continues to decrease, but the current in the circuit still increases. If we could continue to lower the resistance of our appliance to zero (similar to a short circuit), we would discover that the voltage in the circuit would equal zero. (There can be no voltage across a short circuit.) It is interesting to note that the current tends to rise and remain constant at about 700–800 milliamps. (See Figures 2-7 and 2-8.)

From a practical standpoint, a solar cell should be operated between the extremes of an open circuit (zero current, maximum voltage) and short circuit (zero voltage, maximum current) arrangement, since neither of these conditions will enable the cell to produce enough electrical power to efficiently operate our products.

PV cell engineers have calculated a voltage at which the power output of a solar cell or module (a group of cells) is maximized. This particular voltage is referred to as the *maximum power voltage*. Associated with this voltage is a corresponding current, called the *maximum power current*. Together they define the *maximum power output* of a photovoltaic cell, panel, or array.

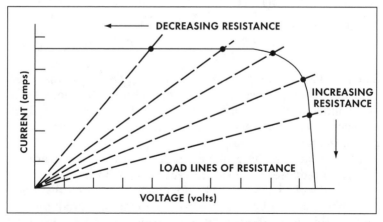

Figure 2-7. Effect of resistive load on cell operating point. As the resistance of a load in a circuit increases, the terminal voltage of a solar cell increases. When the load's resistance decreases, the current in the circuit increases.

Engineers would like to operate solar cells—as well as modules and arrays (see information under the subheading The Basic System, on the following page)—near their maximum power point at all times. However, this energy point changes continually with the changing intensity of the sunlight.

On any particular day, a solar cell might be exposed to the maximum value of sunlight at the earth's surface, about 1,000 watts/meter². In this situation, the PV cell would produce its maximum power. What would happen to the cell's energy output if we reduced the sunlight, as would happen on a cloudy or overcast day? It might appear that PV cells are not very useful on days with limited sunlight, but scientists have discovered that cells

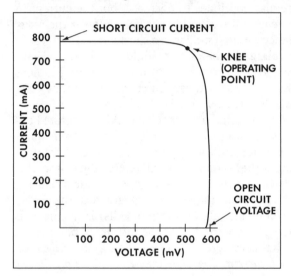

Figure 2-8. Solar cell I-V curve. In practical applications, solar cells are not operated in the open-circuit condition (zero current, maximum voltage) or short-circuit arrangement (maximum current, zero voltage). The cell's efficient operating point occurs at the knee of the curve.

function even when light is reduced! Research indicates that over a broad operating range the *voltage* of a silicon cell is not significantly influenced by the intensity of sunlight. It is not until light becomes very limited—about 8 percent of the maximum sun level—that the voltage drops drastically.[9]

The *current* of a photovoltaic cell does not behave similarly to voltage in reduced light. The current varies with the amount of photons absorbed into the cell. If sunlight is reduced there are fewer photons absorbed into the N-layer and the cell produces less current. However, scientists have discovered that the current is directly related not only to the intensity of sunlight but to the *surface area* of the cell as well. If a cell is made larger, it will absorb more photons and the current will increase, and if it is made smaller, it will produce less current. It is interesting that the cell's voltage changes very little regardless of the cell's size. Thus, solar cells, when operated properly will produce substantial power even in low-light-level conditions.

The power output of solar cells will vary from zero in the morning to the watts peak at noon, and then it will decrease as evening arrives.[10] Engineers realize that these devices can serve as useful electrical generators only when the total PV system is correctly designed.

The Basic System

A small silicon solar cell will produce .45 volts of electricity when exposed to sunlight at midday. Since the size of the cell determines its current rating, values will vary from about .3 amperes to 2 amperes. Certainly, this would not be sufficient electrical energy to power a large-scale product such as a solar electric home. How do engineers take a small photovoltaic device and use it in applications that require a substantial amount of electricity?

Similar to batteries, solar cells can be connected in a variety of circuit designs to increase the voltage, current, or both. If a great amount of voltage is needed for a product, the PV cells can be connected into a *series* circuit (Figure 2-9). In a series arrangement, the *total voltage* of the system is equal to the sum of the voltage of the individual cells.

In this case, the voltage equals 1.35 volts. The current in a series circuit equals the current output of a single cell. In the illustrated circuit, the total current equals 2 amperes.

Some products, such as motors and air conditioners, may draw a large amount of current during their operation. To increase the current that is available in a circuit, solar cells can be grouped into a *parallel* format (Figure 2-9). When arranged in a parallel circuit, the total current equals the sum of the current of the individual cells. (As shown in Figure 2-9, the total current equals 6 amperes.) The voltage in the parallel system equals the voltage of a *single* PV cell.

If an increase in voltage and current is needed, engineers can connect the individual PV cells into a *series/parallel* arrangement (Figure 2-10). In this format, the voltage is determined by the number of series groups of cells, and the current is based on the number of parallel cells in each group.

The constituent element of a photovoltaic system is the indi-

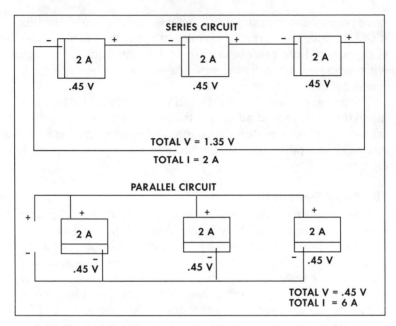

Figure 2-9. When solar cells are electrically connected to a series circuit, voltage equals the sum of the individual cell voltages. In a parallel circuit, current equals the sum of the individual cell currents.

vidual solar cell. Combining a few of these cells into a series, parallel, or series/parallel circuit can increase the supply of voltage and current. Yet, many applications, such as refrigeration, air-conditioning systems, heating units, and lighting for residential and commercial use, draw an enormous amount of voltage and cur-

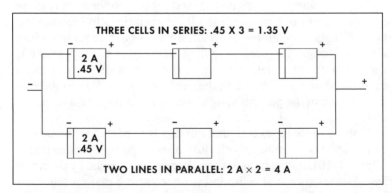

Figure 2-10. Solar cells can be connected in a series/parallel arrangement to increase the total voltage and current output.

Photo courtesy of Richard Embery, Photocom Inc.

Figure 2-11. Kyocera Corp. produces Duravolt™ modules, which are virtually unbreakable and especially suited to remote power applications. The models illustrated are 12.5 W, 25 W, and 50 W designs.

rent—much more electricity than a few PV cells can generate. To meet these high power requirements, engineers have designed *photovoltaic modules* (Figure 2-11).

The PV module is typically composed of 40 to 200 electrically connected solar cells.[11] The cells used in contemporary modules are high-efficiency units (14 to 15 percent efficient) and will last about 20 years. The PV cells are mounted internally to a variety of materials. The Kyocera Company laminates its solar cells between a tempered glass cover and an ethylene vinyl acetate poltant with an aluminum foil backing to provide protection against the environment.[12] The Siemens Company, another major producer of PV modules, electrically matches cells and laminates them in a multi-layered polymer system also using ethylene vinyl acetate.[13] When covered with low-iron-content glass, the cells are protected against salt air, dust, and other environmental influences (Figure 2-12).

The laminated cell system is surrounded by a rectangular, anodized aluminum frame to provide structural strength. The frame is used for mechanically mounting a module in place on a rack, roof, or other support. Located on the back of each unit is a junction box designed for electrical installation into the solar array system.

The size, weight, and energy output of solar modules depend on the particular energy outputs of the units. A PV module may be as small as 10" x 20", weigh only 4 pounds and produce about 11 watts. A larger module will deliver considerably more power, 63 watts; however, its size will increase to 17" × 17", and its weight will increase to about 16 pounds.

Engineers can electrically connect these modules in series or series/parallel to provide dc electricity in a desired voltage/current range for high energy needs, such as a utility grid system for

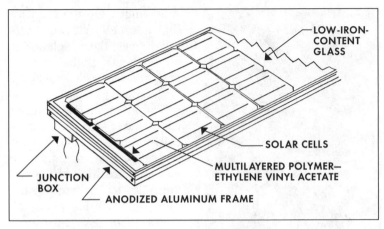

LOW-IRON-CONTENT GLASS

SOLAR CELLS

MULTILAYERED POLYMER—ETHYLENE VINYL ACETATE

JUNCTION BOX

ANODIZED ALUMINUM FRAME

Figure 2-12. Photovoltaic modules manufactured by Siemens Solar Industries consist of high-efficiency solar cells laminated between a multilayered polymer backsheet and layers of ethylene vinyl acetate. The module is moisture resistant, environmentally safe, and ultraviolet stable. Diagram courtesy of Siemens Solar Industries.

a home. When several modules are linked together, they are called a *photovoltaic array* (Figure 2-13). How many modules are needed to create a solar array? A small array may consist of four modules, and a large array may comprise several hundred units (Figure 2-14). The number of PV modules really depends on the operating requirements of the system.

Figure 2-13. A small ground-mount stand by Siemens Solar Industries will hold two to four modules in a solar array. The modules can be connected together to produce the required voltage and current outputs (dc) for the given applications.

The photovoltaic array is finding uses in solar electric home designs, in utility systems providing power for communities, and in remote geographic regions where conventional power lines do not exist. (See Chapter 1 for a description of the applications of photovoltaic modules, arrays, and systems.)

In 1983, Siemens Solar Industries constructed a 6.5 megavolt utility power station in California using photovol-

Figure 2-14. A large-scale photovoltaic array constructed by Siemens Solar Industries. This Caltrans installation provides electricity in a remote region.

taic cells, modules, and arrays. These large-scale megawatt power plants attest to the reliability, performance, and visibility of photovoltaics.

Notes

[1] Infrared wavelengths of photons possess low energy and are not efficient generators of electricity. Ultraviolet (shorter) wavelengths generate excess energy, which is transformed into heat. The middle range of wavelengths converts effectively into electricity.

[2] A good conductor of electricity is a material that allows its outer shell electrons to move easily from one atom to another.

[3] Each photon affects only one electron, and any excess energy is released as heat.

[4] Although the "hole" does not move, scientists refer to the process as "electron hole movement."

[5] Charles Wardel, "Making A Solar Cell," *Fine Homes Builder* 72 (March 1992), 75.

[6] *How a Solar Cell Works* (Billerica, MA: Mobile Solar Energy Corporation, 1992).

[7] Edward Edelson, "Solar Cell Update," *Popular Science*, June, 1992, 96.

[8] The efficiency of solar cells has steadily increased during the past 10 years, from about 7 percent to 16 percent.

[9] Edward Roberton, *The Solar Guide to Solar Electricity* (Frederick, MD: Solarex Corporation, 1979), 29.

[10] Watts peak is a common unit of measure that identifies the power output of a solar cell at midday under a clear sky.

[11] Miles Russell, *Residential Photovoltaic Systems Handbook* (Reading, MA: Sundance Publ., 1984).

[12]Barbara Peppriell, ed., *Photocom, Inc. Solar Electric Power Systems Design Guide* (Scottsdale, AZ: Photocom, Inc., 1991).

[13]*High Efficiency Solar Electric Modules* (Camarillo, CA: Siemens Solar Industries, 1991).

CHAPTER THREE

Materials and Processes

THE promise of a clean, renewable source of electricity derived from the sun is a reality. Solar electricity is making its way out of the laboratory and moving into applications for utility companies and consumers. Scientists and engineers worldwide are working on new materials to more efficiently convert sunlight into electricity. They are also developing mass-production methods that will make electricity produced from solar cells cost competitive in terms of cost with that produced by conventional sources of energy.

Although many photovoltaic technologies exist, most approaches to building solar cells are based on the use of silicon. In part, silicon has been extensively used in the photovoltaic process because it is abundantly available in the form of silica, a high-grade sand. Although easily obtained, silicon must be refined to remove impurities that would interfere with its proper functioning in a solar cell. Until recently, the process of purifying silicon was expensive. Now, however, the use of new technologies has reduced cost while maintaining the appropriate purity level of the material.

Silicon: The Czochralski Process

To use silicon in the photovoltaic process, and to package the material into cells, modules, and arrays, scientists devised a method of growing a single ingot of the material that can be sliced into "wafers." The process involves using purified silicon in a block form, a form actually comprising many small crystals of silicon that are often referred to as *polycrystalline silicon*. In the Czochralski process (Figure 3-l), engineers melt the polycrystalline silicon in a temperature-controlled crucible. A seed crystal of the silicon is then dipped into the crucible. The single crystal is then drawn up from the molten silicon very slowly, allowing the molten material to solidify in a way that reproduces the crystalline design. This process creates silicon ingots 6" in diameter and 36" long.

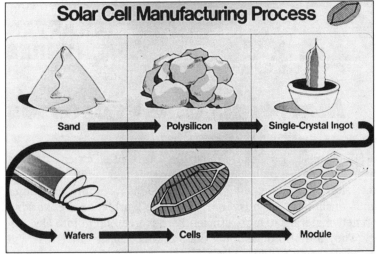

Solar Cell Manufacturing Process

Sand ▶ Polysilicon ▶ Single-Crystal Ingot

Wafers ▶ Cells ▶ Module

Photo courtesy of Siemens Solar Industries, Camarillo, CA

Figure 3-1. A simple view of module fabrication.

After undergoing a cooling process, the hardened cylinders of silicon are sliced into thin wafers using special diamond abrasive saws. The saws produce silicon wafers that are 1/100" thick.

The photovoltaic effect produced by the wafers is achieved by adding the dopants boron and phosphorus to the silicon. The boron is added during the molten drawing stage. Phosphorus is added by placing it with water in a furnace and allowing a small amount of the resulting vapor to penetrate the silicon wafer's surface. Metal contacts are added to each side of the solar cell and the entire wafer is then covered with a protective coating of silicon.

Manufacturing cylinders of pure single-crystal silicon and slicing them into solar cell wafers (Figure 3-2) is an expensive process. Part of the procedure requires handwork, which adds to the cost of production. In addition, some of the silicon is wasted during the sawing process. It is estimated that half of the expensive silicon bar is lost as dust during the sawing process.[1]

Although growing an ingot of pure single-crystal silicon has been the traditional material and method used to create a photovoltaic cell, scientists have also experimented with processing other types of silicon. Polycrystalline silicon (also known as polysilicon) is cheaper to produce than pure crystals, and it will function as a

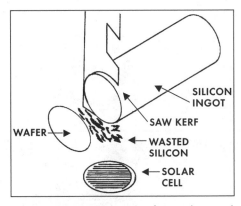

Figure 3-2. Production of a single-crystal silicon cell

photovoltaic material. Polysilicon does not have to be grown as a single crystal but can be formed as a thin ribbon of material or even transformed into silicon vapor and deposited on a thin surface of another material.

To create polysilicon, engineers cool molten silicon very slowly. Unfortunately, the material that results from this process may consist of many tiny crystals of silicon. Each crystal has a "grain" boundary between itself and the next crystal that causes electrical shorts in the material. These electrical shorts inhibit conversion of sunlight into electricity, which means that polycrystalline silicon has had a lower conversion efficiency than single-crystal silicon.

Some companies, such as the Japanese firm Sharp, have succeeded in reducing the electrical shorts in polysilicon. Researchers at Sharp have reported conversion efficiencies of 16.4 percent and hope to achieve greater efficiency in the near future.[2]

Edge-Defined Film-Fed Growth Process

A promising technology for producing silicon that is nearly single-crystalline in structure involves the edge-defined film-fed growth process (EFG). This revolutionary technology produces polycrystalline silicon in sheet form. It is adaptable for mass production of photovoltaic cells and is less expensive than the single-crystal growth process.

Mobile Solar Energy Corporation is one of the leading companies producing sheet silicon by the EFG method. The company forms silicon sheets directly—a thin sheet emerges directly from the molten silicon, with no significant loss of material due to the con-

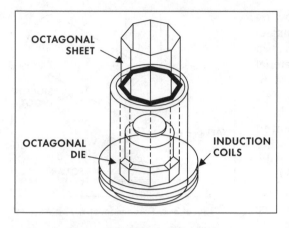

Figure 3-3. EFG crystal growth

Figure 3-4. Mobile Solar's EFG crystal growth furnaces. Hollow, eight-sided tubes are grown from molten silicon.

ventional sawing methods. The process uses a graphite die that is filled with molten silicon (Figure 3-3). The liquid silicon rises by capillary action through a very narrow die slot and forms a curved surface at the top of the die. A seed crystal of silicon contacts the molten silicon at the liquid's curved surface, forming a bridge between the die and the seed crystal.

As the liquid silicon is drawn up from the die, it is cooled, and at the same time a crystal grows from the seed downward to the original contact position. The silicon solidifies in sheet form. Capillary pressure pushes the molten silicon column up the die to continually feed the crystal's growth. The size of the sheet is defined by the die-top edge—hence the name *edge-defined film-fed growth process*.[3]

The EFG process is very stable and eliminates the need for technicians to rigorously control the system in terms of temperature and speed. Consequently, it is possible to automate the growth process of the silicon sheet (Figure 3-4).

Today, Mobile Solar Corporation uses an eight-sided tube with a 4" width to develop its solar cell material. This octagonal tube produces 32"-wide silicon sheet in 15' lengths. A high-power laser cutting system is used to saw the material into thin wafers. The laser cutter reduces wafer breakage and less silicon is wasted than in conventional processes.

Photo by R. Schliepman

Figure 3-5. A technician inspects a module as it is removed from the laminator. Mobil Solar's 4' x 6' module is designed specifically for utility grid-connected applications.

The edge-defined film-fed growth process can produce photovoltaic cells with a sunlight-to-electricity conversion efficiency comparable to other crystalline cells. A conversion efficiency of 13 percent has been achieved in pilot production and 15 percent in the laboratory. The next step in addressing the efficiency barrier is to overcome the interactions between missing or extra silicon atoms and electrically charged chemical impurities in the material.

Recent advancements such as doubling cell size, reducing the process sequence, and lowering the cost of the automated manufacturing process will further reduce the cost of producing photovoltaic cells (Figure 3-6).

Thin-Film Photovoltaic Cells

The problem of building inexpensive photovoltaic cells may have its solution in amorphous materials. Amorphous silicon is not built on a crystalline structure—its atoms are randomly distributed. Consequently, a small amount of amorphous silicon, such as a thin layer, will convert sunlight into electricity.

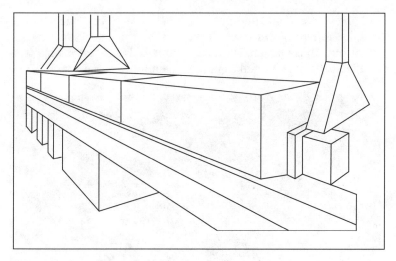

Figure 3-6. Automated cell manufacturing

Since production of amorphous solar cells does not require the usual crystalline processes, it is less expensive than the standard silicon ingot cells. Hair-thin layers of amorphous silicon can be deposited on thin sheets of steel or plastic to make them a versatile energy source for small products such as watches and calculators.

Amorphous silicon cells have had one major problem: a low sunlight-to-electricity conversion efficiency. Initially, the thin-film amorphous material had only a 6 to 8 percent conversion rate, which limited its usefulness to serving as an energy source for small calculators and watches.

Today, a new process, developed by Energy Conversion Devices (ECD) creates roll-to-roll thin-film amorphous silicon solar cells that have a conversion rate of about 14 percent (Figure 3-7). The inventor of the process, Stanford Ovshinsky, places a thin film of amorphous silicon (rather than costly crystalline silicon) on a long roll of stainless steel sheets. The silicon is deposited on the steel sheet as it moves through deposit chambers. Energy Conversion Devices' machines can produce 2,500-foot rolls of solar cells.

The photovoltaic cells produced by this process possess a higher conversion efficiency than typical amorphous silicon cells. These solar cells are made into *triple-cell* structures. Three transparent layers of thin-film amorphous silicon capture a wider range of light wavelengths than a single-layer cell. The cells also use amorphous silicon alloys and an amorphous silicon germanium alloy.

This process may reduce the cost of solar-generated electricity

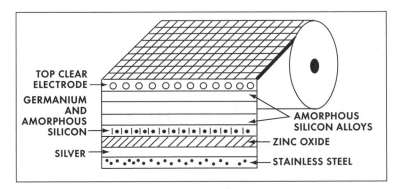

Figure 3-7. Roll-to-roll solar cells

to about 10¢ per kilowatt-hour within 10 years. Ovshinsky and his ECD firm have already sold a two-megawatt-per-year solar production system to the Russian energy company Kvant.[4]

The lure of thin-film photovoltaic cells is related to the variety of inexpensive materials that will convert sunlight into electricity. Cells made of cadmium telluride have produced conversion efficiencies in the laboratory of 14 percent; although this is not an extremely high level, the low cost of the cells produced tends to offset their lower efficiencies.[5]

Thin-film materials have also been developed into *tandem* cells (Figure 3-8). These cells have layers of materials that respond to different portions of the solar spectrum. One layer absorbs high-energy photons of sunlight while the second layer absorbs and is activated by photons of lower energy levels. Boeing Corporation has developed a two-layered tandem cell based on gallium materials. Using gallium arsenide and gallium antinomide, Boeing has created a cell that converts 37 percent of sunlight into electricity.

An interesting application of thin-film technol-

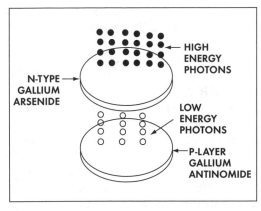

Figure 3-8. Tandem cell

ogy is being developed by Germany's Flaschglaff Solar Company. Flaschglaff's approach involves integrating thin-film solar cells into window glass creating a power-generating glass system. This would enable architects to use windows as electricity-generating devices for their structures.

In the United States, the Mazda Corporation now offers a solar-powered ventilation system as an option in one of its luxury car models. The system uses solar cells integrated into a glass sun-roof panel that generates electricity from sunlight both to power the ventilation system when the car is idle and to trickle charge the car's battery.

Although photovoltaic cells made from thin films tend to degrade over time when exposed to sunlight, the outlook for the future suggests that this problem, as well as the lower conversion efficiencies, will be solved. The promise of thin-film technology lies in the mass production of these cells to bring cost competitive sunlight-produced electricity into reality.

Spheral Solar Technology

In 1992, Texas Instruments Corporation and Southern California Edison announced a co-development agreement to explore a new photovoltaic technology, *spheral solar electricity*. This new technology may result in achieving two significant goals: (1) producing electricity from sunlight in an inexpensive manner and (2) developing a low-cost manufacturing process to produce the new photovoltaics.

Texas Instruments' approach is based on neither thin-film technology nor crystalline silicon cells. Spheral technology uses low-cost metallurgical-grade silicon. This silicon differs from standard-grade silicon in that it has a lower level of purity. This translates into lower cost, about $1 per pound as compared with solar-grade silicon costs ranging from $5 to $7 a pound.[6]

The spheral solar process begins with silicon that contains impurities. To remove these impurities, "chunks" of silicon are sent through an open-air furnace that grows an oxide on the silicon's surface. The material is then sent through a melt furnace where the silicon oxide layer pulls the melting silicon into a spheral shape.

In the third phase of the process, the silicon is placed in a freeze mode. As the silicon freezes, impurities are pushed to the surface of the material. The impurities can then be ground away, and the process can be repeated until the desired level of purity is achieved. The silicon is now known as P-type material. (See Chapter 2 for a discussion of P and N layers.) The addition of an N layer creates a functional cell.

Next, the silicon spheres are pressed into a sheet of aluminum foil, similar to the kind used to wrap food. The aluminum foil is embossed and etched, creating a sheet 12" x 4". The 54,000 silicon spheres are loaded onto the sheet and metallurgically bonded to the aluminum.

The front of the foil is in contact with the N layer of the silicon and the back of the foil is etched to allow the P silicon to be connected to a second piece of foil. A second layer of foil is added to this back side and the entire system is pulled together using a vacuum process (Figure 3-9).

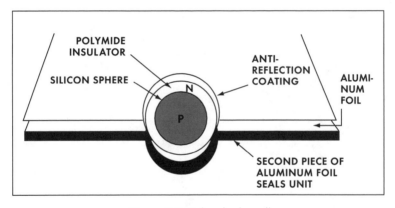

Figure 3-9. Spheral solar cell

The silicon sphere and the aluminum are protected against electrical short circuiting by a coating of a thin polymer material. Finally, the cells are coated with an antireflective material.

The completed photovoltaic cell is a flexible unit that needs little conventional wiring, and cells can be connected into either a series or parallel circuit to meet a variety of applications. Although the cell's 10 percent conversion level is lower than that of other solar cells, the reduction in cost with respect to the use of lower-grade silicon and mass-production potential may reduce the current cost of photovoltaics from 30¢ to about 15¢ per kilowatt hour.

California will be one of the first sites for application of this photovoltaic technology. Spheral technology will be used in a solar carport that can be used to recharge electric autos whose use should become more common due to a recent legislative mandate calling for more antipolluting vehicles by the year 2000.[7]

Notes

[1]Paul Maycock and Edward Stirewalt, *Photovoltaics: Sunlight to Electricity in One Step* (Andover, MA: Brick House Pub., 1981).

[2]Edward Edelson, "Solar Cell Update," *(Popular Science*, June 1992): 95-99.

[3]Mobile Solar Energy Corporation, *EFG Photovoltaic Technology for Electric Utilities* (Bellerica, MA, 1992).

[4]See note 2 above.

[5]Efficiency achieved by Professor Ting Chu of the University of South Florida.

[6]Eric Graf, "Spheral Solar Technology" (technical presentation, Soltech Conference, 1992).

[7]Mark Uehling, "Silicon Bead Cells," *Popular Science*, (June, 1992).

CHAPTER FOUR

Solar Cells for the Experimenter

CONSTRUCTING projects that are powered by the sun is both enjoyable and exciting. The experimenter has two readily available types of basic cells. The first is a small unit (2 x 4 cm) that provides .55 volts at 300 milliamperes on a clear sunny day (Radio Shack No. 276-124). (See Figure 4-1.)

The second type of cell, the *mini panel*, consists of a photovoltaic cell placed inside a plastic case. The cell's leads are preattached to the unit (Figure 4-2). The mini panel is available in two versions: .45 volts at 400 milliamps (Pitsco No. 08631) and 1.5 volts at 200 milliamps (Pitsco No. 08635).

Solar cells are delicate and must be handled with care. However, if you accidentally crack or chip a cell, do not throw it away, because it will still function to some degree (Figure 4-3). If a small piece of the cell has broken off the edge, the cell will still possess most of its ability to provide energy. If a cell breaks in half, it will still have a small portion of its energy-producing capability.

Photovoltaic cells must have wire leads soldered to their positive and negative terminals. (As mentioned above, encased cells have preattached leads.) The negative terminal of the cell is a silver-colored strip located on the "top" of the cell (the dark blue side). (See Figure 4-4.) The positive terminal is located on the "back" of the unit.

Figure 4-5 shows some of the most basic items you will need to solder the wire leads to the cell's terminals. First, selecting the appropriate wire for the leads is very important. You can quickly and easily solder a thin-gauge wire, such as 30-gauge computer wire, to the cell. This protects the unit from overheating. Applying too much heat to a solar cell can destroy its effectiveness.

Second, you'll need an

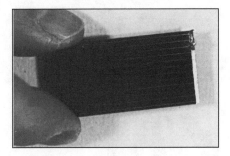

Figure 4-1. A small solar cell, 2 cm x 4 cm. Its output is .55 V at 300 mA. This cell must have positive and negative lead wires attached to its terminals.

Figure 4-2. The solar mini panel. It is encased in plastic and comes with preattached leads.

appropriate soldering iron. A small, lightweight, 30-watt pencil iron is easy to manipulate and will not overheat and destroy the solar cell. (The photo shows a Radio Shack soldering iron type 64-2067.) Third, you will need solder. A standard 60/40 rosin-core solder will work fine. Select a thin diameter for the solder, to avoid depositing an excessive amount of it on the cell (.062 diameter, Radio Shack type 64002, works well). Finally, you will need a simple pair of tweezers to hold the solar cell during the soldering operations.

Allow your soldering iron to reach its maximum temperature before you begin to solder the leads to the solar

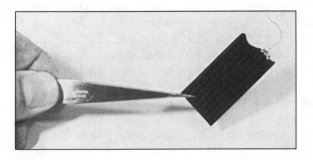

Figure 4-3. A solar cell will still operate even with a small section chipped off.

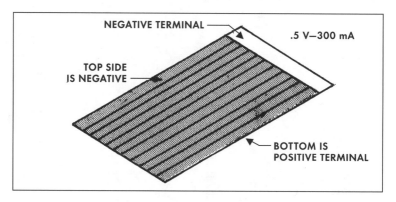

NEGATIVE TERMINAL

.5 V—300 mA

TOP SIDE
IS NEGATIVE

BOTTOM IS
POSITIVE TERMINAL

Figure 4-4. Solar cell connections

cell. Apply a little heat to the cell's negative terminal, then place the wire lead in position. Carefully apply heat from the soldering iron and add a *small* amount of solder to the joint. Try not to over-

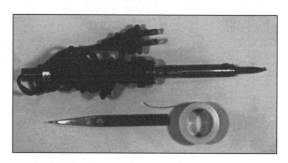

Figure 4-5. A small 40-W soldering iron, a pair of tweezers, and fine-gauge wire are needed to attach the leads to a solar cell.

heat the cell, to avoid damaging its operation. Allow the joint to cool before moving the cell. Next, turn the cell onto its back side. Note that the entire back side is a positive terminal, which means that you can solder your lead to any convenient location.

You can avoid the problem of soldering leads to a cell by purchasing a plastic-encased photovoltaic unit. The advantages of the unit are that the cell is protected by the casing and you don't have to bother with attaching the wire leads. There is, however, a disadvantage of this type of unit: To increase the output you must wire several of the units together. Since each cell is inside a case, the overall physical size of your system may become too large for your project. You'll find the encapsulated cells an excellent choice when you need only one or two solar cells to operate your project.

Project Section

The following projects are designed to illustrate the basic concepts of photovoltaics introduced and discussed earlier in this book. The activities range from low-cost, easy-to-build projects to medium-priced, challenging and operational models. An understanding of electronic circuit construction and a basic knowledge of tools, materials, and processes (drilling, filing, etching, gluing) is helpful in completing the projects. Each project includes detailed, step-by-step instructions, a schematic (where needed), a parts list, and other useful illustrations. Although materials and parts are suggested for each project, the models may be altered to fulfill particular needs or cost objectives. For your convenience, a Directory of Suppliers appears on page 96.

Construction of these projects will make learning about solar electricity more fun and should encourage exploration of other uses of photovoltaics.

CHAPTER FIVE

Solar Sun-Racer Vehicle

INFLUENTIAL states, such as California, have required that a larger percentage of all cars produced be "nonpolluting." As a result, toward the end of the 1990s, several automobile manufacturers will introduce mass-market electric cars to the public.

The solar sun-racer vehicle provides a fun and interesting project for exploring the concept of the electric car. (See Figure 5-1.) This little car is based on a frame made of model brass tubing and hollow brass squares, a standard 1.5 volt hobby motor, model airplane servo gears, a set of model race car wheels, and two 1.25 volt ni-cad batteries. (The solar panel that recharges the batteries is constructed using the plans discussed in Chapter 8's solar flashlight project. (See page 75.))

The sun racer will run approximately 30 to 45 minutes on a full charge. To recharge the car's batteries simply plug the solar panel into the battery socket and place the unit in bright sunlight for approximately three hours.

Tools and glue needed to construct the sun racer:
- K & S tubing cutter
- Dremel motor tool with a cutoff disk or a coping saw
- Soldering iron
- Cyanacrylate glue (such as Super Glue)
- Epoxy glue
- Small flat file
- Small convex-shaped file

Figure 5-1. The solar electric model car has a brass frame and is powered by two 1.25 V rechargeable ni-cad batteries. It is recharged using the panel whose construction is described in Chapter 8.

Construction Procedure

1. Carefully review all photographs and drawings before beginning construction of the model.

2. Cut two pieces of 3/32" square hollow brass to a length of 6" (Figure 5-2). These will serve as the side frames of the vehicle. You'll find that a Dremel motor tool with a cutoff disk works nicely for this process.

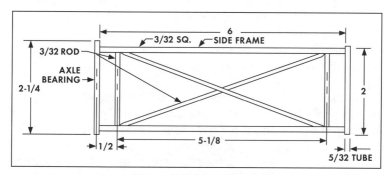

Figure 5-2. Frame layout

3. The side frames are connected to the front and rear axle bearings (Figure 5-2). The axle bearings are made from 5/32" diameter brass tubing. The front bearing is 2-1/4" long and the rear bearing is 2" long. Cut the tubing with a K & S tubing cutter or your Dremel motor tool.

4. Clean the ends of the tubing with a file, because burrs will prevent the axles from turning smoothly.

5. Using a small convex-shaped file (e.g., pattern file), shape the ends of the square side frames until they mesh flush against the tubular axle bearings.

6. Using Figure 5-2 as a template, lay the side frames and axle bearings in position. Mark the locations where the pieces will be joined. Glue the two side frames to the front bearing using a Super-Glue-type product. Then, glue the rear axle bearing into place, again using Super Glue.

7. Let the glue dry for a few minutes.

8. To reinforce the basic frame's joints, apply a small amount of epoxy glue (use the two-part type of epoxy) to the joints. Allow the epoxy to dry for approximately 30 minutes.

9. Next, add the support rods as shown in Figure 5-2. The support rods are made from 3/32" brass rod. The first support rod is glued in place 1/2" from the front axle bearing. The second rod is glued 5-1/8" from the first support rod. First, fix it in place with Super Glue, then strengthen the joint with epoxy.

10. Add the cross members as shown in Figure 5-2. These are three pieces of 3/32" diameter brass rod. One piece should run from the first support to the second. Sand the edges at an angle to fit flush with the side frames. The other two pieces should run from the supports to the center of the single cross brace you just installed. Super Glue the pieces, then apply some epoxy glue for added strength. Set the completed frame assembly aside to dry.

11. Cut a piece of 1/8" diameter brass rod to a length of 2-13/16". This piece will serve as the front axle. File off any burrs that might obstruct the axle's turning. Insert the axle in the front bearing.

12. Place the two front wheels (model racing car wheels or Pitsco wheels—see the parts list) on the axle. Make sure the axle turns freely in the front bearing. Epoxy bond the wheels to the axle.

13. Next, cut a piece of sheet brass to a finished size of 1-15/16" x 2-3/4". This piece will serve as the "component plate" (Figure 5-3). The motor and batteries will be mounted on this piece of brass.

14. Use Super Glue to attach two N-size battery holders to the front of the plate. The battery holders on the prototype model were set flush to the edge of the brass piece (Figure 5-1).

15. The motor is attached with Super Glue to the "rear" edge of the component plate. *Note that the motor must be correctly aligned to the rear axle spur gear before it is glued in position.*

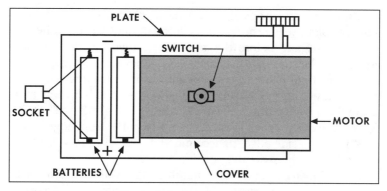

Figure 5-3. Component layout

16. Carefully review Figure 5-4, which covers the motor and axle spur gear system. Select two gears from your gear sets. (See the parts list.) The prototype used a motor gear and spur gear ratio of approximately 1:1. Although this provides low torque, it does produce a fast-running car.

Figure 5-4. Select two gears that will determine the speed of the car. The gears must mesh smoothly before the motor mount plate is bonded in position.

17. Cut a rear axle to 1/8" x 1-13/16". (Your rear axle may have to be longer, depending on the particular gears you have selected.) Slide the axle through the bearing, then slide your spur gear on the axle. Place both wheels on this rear axle and adjust the gear and wheels so the car rolls freely.

18. Place the gear you have selected for your motor gear on the motor's shaft. Place the motor mount plate on the frame, and hold the motor on the plate. Slide the motor and plate towards the rear axle until the motor's gear meshes smoothly with the axle gear. (See Figure 5-4.) Mark the motor's position on the plate so that you can glue it in the same position.

19. Using Super Glue, carefully attach the motor to the component plate. Before the glue sets, slide the plate and motor until the motor's gear once again correctly meshes with the spur gear. Make any necessary minor adjustments before the glue dries.

20. Place a few drops of Super Glue along the edges of the components plate while the motor is correctly meshed with the axle spur gear. Again, before the glue dries, move the car back and forth to make sure that the transmission meshes smoothly.

21. The final phase of construction involves making the coverplate and wiring the model. (See Figures 5-3 and 5-5.) Make the cover plate from a thin piece of shim brass 1-3/4" x 3-1/2". Drill a 3/16" hole in the cover, located 1" back from the front edge and 1-1/4" from the side edge, to accommodate the spst off-on switch.

22. Mount the switch in position.

23. Following the wire diagram in Figure 5-5, wire and solder the entire electrical system. Make sure you observe the correct polarity of the components. The socket used in the system should be a "mate" to the "plug" used on the solar charger unit. (See Parts List and plans for the solar charger unit in the solar flashlight project in Chapter 8.)

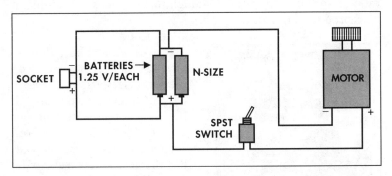

Figure 5-5. Schematic

24. Once the system has been wired, you can attach the cover plate to the model. Slide one end of the cover plate between the rear support and the rear axle. Force the brass over the motor and press the front edge into the grooves on the ends of the battery holder. Epoxy bond the cover plate in position.

Charging Procedure

1. Lay the solar panel charging unit on the front of the car and plug the unit into the battery's socket (Figure 5-6).

2. Remove one of the batteries. Set the unit in bright sunlight for two and a half hours. The battery will be charged at the end of this time. Remove the charged battery and charge the second battery following the same procedure.

Figure 5-6. The solar panel fits nicely on the front of the car for recharging. Simply plug the panel in the battery's socket and set the unit in the sun for three hours.

3. When you are charging the unit, make sure that the spst switch is in the OFF position.

4. After the unit is charged, unplug the solar panel and turn the switch to its ON position. Gently set the car down—and watch it run! (The prototype model ran 45 minutes on a set of fully charged batteries.)

Note: Again, Chapter 8's solar rechargeable flashlight project for fully illustrated plans and construction procedures for the solar charger panel.

PARTS LIST		
Item	**Description**	**Quantity**
Side frame	3/32" × 6", 3/32" square hollow brass; K&S Engineering Co. (Found in most local hobby shops)	2
Front axle bearing	5/32" × 2-1/4", 5/32" dia tubing; K&S Engineering Co.	1
Rear axle bearing	5/32" × 2", 5/32" dia brass tubing; K&S Engineering Co.	1

Item	Description	Quantity
Front and rear support rods	3/32" dia brass rod; K&S Engineering Co.	1, each rod
Cross brace support rods	3/32" dia brass rod; K&S Engineering Co.	1, each support
Front axle	1/8" × 2-13/16", 1/8" brass rod; K&S Engineering Co.	1
Rear axle	1/8" × 2-13/16", 1/8" brass rod; K&S Engineering Co.	1
Motor mount plate	1-15/16" × 3/4", brass sheet; K&S Engineering Co.	1
Motor	Mabuchi 1.5 V hobby motor, #08632, Pitsco, Radio Shack hobby motor #273-223, 1.5 - 3 V	1
Battery	1.25 V ni-cad N-size battery; Radio Shack #23-121	2
Battery holder	N-size, single type; Radio Shack #270-405	2
Switch	spst, submini switch; Radio Shack #275-645	1
Wheels	1-3/8" dia with a 1/8" axle hole; local hobby shops or Pitsco wheel #77780	4
Connector plug and socket	W.S. Deans Co. plug and and socket connector pin set; local hobby shop model airplane suplies, or radio-controlled	1 set
Motor cover gears	Model airplane servo gears or model radio-controlled car gears such as Troxxas Performance Parts #2011; local hobby shop or Pitsco, gear set #76862	See text

CHAPTER SIX

Solar-Powered Rover Robot

INDUSTRIAL robots currently perform a variety of tasks, such as assembling cars, organizing and shipping materials, and even security patrol operations. Today, robots are often battery powered. However, as they become capable of performing outdoor functions, they will lend themselves to solar electric energy.

The rover robot described here is a small working model that is powered by a series of photovoltaic cells (Figure 6-1). When exposed to sunlight, it will travel along in one direction. You can reverse its direction by simply reversing the plug connecting the solar panel to the motor.

The rover's design is based on four systems: (1) a solar electric panel, (2) a micromotor, (3) a drive transmission, and (4) a frame body with wheels.

The solar electric module has three to six .55 V, .3 A photovoltaic cells. The robot will move along in bright sunlight powered by a solar panel with a minimum of three cells wired in series (1.65 V, .3A). The prototype rover has six cells wired in series (3.30 V, .3A) and travels along at a fairly quick pace in bright sunlight. Since the solar electric panel is held in place by tape and can be removed from the motor, you can experiment with solar panels that produce different electrical outputs—3 cells, (1.65 V), 4 cells (2.2 V), or 6 cells (3.3 V).

A *micromotor* is used to drive the rover. (See the parts list.)

Figure 6-1. The solar electric rover. The frame consists of sheet brass held together with screws, spacers, and nuts. A photovoltaic panel supplies power to a small micromotor that drives the transmission.

These small motors are very sensitive, and they operate on a minimal amount of current and voltage. Micromotors are available with built-in or attachable gear heads that increase the motor's torque. Although the rover project used a built-in gear head, the type of gear head and the motor's size are *not* critical, since the motor can be mounted on a spacer to adjust its position in relation to the drive transmission.

The drive transmission (Figure 6-2) used in the project is a servo gear set used in radio-controlled model cars and airplanes. These are inexpensive plastic gears that offer a variety of transmission designs.

The frame and wheel assembly of the rover are made from sheet brass, 2" #6-32 screws and nuts, brass tubing, brass wire, and four

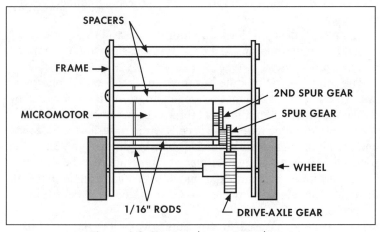

Figure 6-2. Frame and transmission layout

Figure 6-3. The rover's photovoltaic panel is attached with double-sided tape. The rover operates on a minimum of three solar cells wired in series.

model airplane wheels. The frame could be made from other materials such as thin sheet aluminum, thin Plexiglas, or a sturdy piece of matte board.

The rover is a fun project, and with a little experimentation and imagination you may be able to modify it into a sophisticated robot that performs different functions.

Construction Procedure

1. Carefully review the photographs (Figures 6-1 and 6-3), the schematic drawing (Figure 6-4), and the side frame layout drawing (Figure 6-2), before you begin construction.

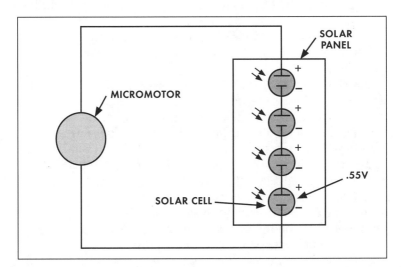

Figure 6-4. Schematic

2. Cut two pieces of .032 sheet brass to a finished size of 1-5/8" x 2-5/8" as shown in Figure 6-5. (You can substitute Plexiglas or aluminum, if desired.)

3. Lay out two axle holes on each side frame. The holes are 1/8" up from the bottom edge and 1/4" in from the ends of the frame, as shown in Figure 6-5.

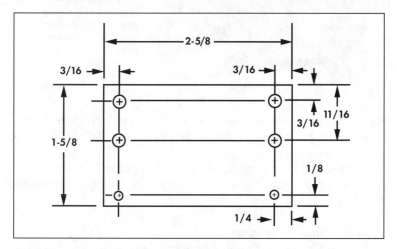

Figure 6-5. Frame layout

4. Using a hammer and a punch, make a light punch mark at the center of each hole. The punch mark will prevent your drill bit from skidding on the brass's surface.

5. Secure one side-frame piece in a vise, and using a 3/32" drill bit, drill each axle hole. Next, drill the axle holes in the second side frame.

6. Using a piece of 3/32" diameter brass rod, check the fit of the axle in the holes. If the axle material fits too tightly, enlarge the holes with a small round file. The axle material should turn freely.

7. Lay out four mounting holes on each side frame as shown in Figure 6-5. The top mounting holes are 3/16" in from the edge and 3/16" down from the top. The lower mounting holes are 3/16" in from the edge and 11/16" down from the top.

8. Make a light punch mark at the center of each hole.

9. Using a 9/64" drill bit, drill each mounting hole. File off any burrs remaining on the frames from drilling operations.

10. Cut four pieces of 3/16" diameter brass tubing to a length of 1-5/8". These four pieces serve as the frame's spacers (Figure 6-2). A small tubing cutter will make a clean cut in the soft brass. A

Dremel motor tool with a cutoff disk also makes an accurate cut in the tubing.

11. Using four 2" long #6-32 screws and nuts, assemble the frame system (Figure 6-2).

12. Cut two axles from 3/32" brass rod. Each axle is 2-9/16" long. File the ends of the axles and check their fit in the frame assembly.

13. Enlarge the small hole in the large servo gear to 3/32" diameter. (The servo gear is 11/16" diameter, the gear with the largest diameter in the set.)

14. Slide the large servo gear onto one of the axles. Figure 6-2 shows its position. (Note that this is a *trial* position, based on the particular size micromotor you select.) Do not glue the gear in place at this time.

15. Select a mating gear from the servo set. Drill a 1/16" hole in the gear. Slide the gear onto a piece of 1/16" brass rod. The gear must turn freely. If it is tight, enlarge the hole in the gear with a small file or a 5/64" drill bit.

16. Once the gear turns freely on the 1/16" rod, cut the rod to a length equal to the width of the inside dimension of the frame.

17. Check the fit of the rod and spur gear—it should fit tightly against the frame's sides, with the spur gear meshing into the drive gear on the axle (Figure 6-6).

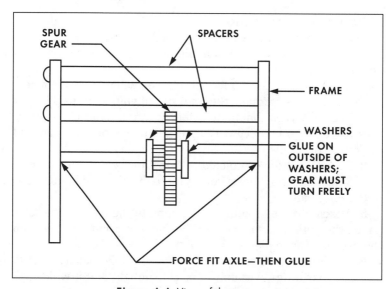

Figure 6-6. View of the spur gear

18. Remove the rod and spur gear. Slide small washers onto the rod, one on each side of the spur gear. Then, set this unit aside.

19. Using epoxy glue, bond two 1" diameter model airplane wheels onto the axle with the drive gear. Bond two wheels onto the remaining axle, making sure the wheels do not rub against the frame's sides.

20. Slide the 1/16" rod with its spur gear and washers inside the frame, meshing and aligning the spur gear with the drive gear. Make sure that the *spur gear rod is straight* in alignment with the *axle gear rod*. Bond the rod in place with epoxy glue. Set the unit aside to dry.

21. Glue a second spur gear onto the micromotor. The gear should be identical to the gear mounted on the frame system.

22. Holding the micromotor with gear against the inside of the frame and meshed with the previously mounted spur gear, realign both the spur gear and the drive gear so they mate properly. Refer to Figure 6-2 and the photograph of the model.

23. Glue the micromotor against the side of the frame using Super Glue. Then, run a bead of epoxy glue around the motor's edge for added security. For additional support you may glue a piece of 1/16" rod across the motor to each side of the frame.

24. Solder a two-pin connector plug to the motor's lead wires. Set the unit aside to dry.

Solar Panel Construction

The final phase of assembly involves the construction of the solar panel. This unit requires a minimum of three solar cells (1.65 volts) to power the rover.

1. After selecting the number of cells you will use, lay them next to each other to determine the total size of your solar panel.

2. The solar cells should be soldered into a *series* circuit (Figures 6-4 and 6-7). (Review Chapter 4 for tips on handling and soldering solar cells.)

3. Epoxy bond the completed cells to a piece of matte board.

4. Solder a mating socket connector to the plus and minus leads of the solar panel.

5. Attach a piece of double-sided tape to the back of the solar panel, then press the panel onto the frame. It should fit between the two side frame pieces.

6. Plug the motor's pin connector into the solar panel's socket.

Place your little rover unit in bright sunlight and it will "putt-putt" right along!

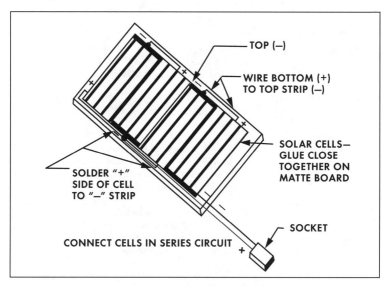

TOP (−)

WIRE BOTTOM (+)
TO TOP STRIP (−)

SOLAR CELLS—
GLUE CLOSE
TOGETHER ON
MATTE BOARD

SOLDER "+"
SIDE OF CELL
TO "−" STRIP

SOCKET

CONNECT CELLS IN SERIES CIRCUIT

Figure 6-7. Photovolatic panel

You can also do the following experimentation:

● Create different solar panels using 4 or 6 solar cells and note the changes in speed and torque.

● Design a control tether for making the unit go forward or reverse.

● Design the unit to perform a fun and interesting task.

PARTS LIST		
Item	**Description**	**Quantity**
Side frames	032 brass sheet; 1-5/8" × 2-5/8"	2
Spacers	3/16" dia brass tubing, 1-5/8" long; K&S Engineering Products	4
Screws nuts	#6-32 screws and nuts, 2" long	4
Servo gear set	Troxxas Performance parts, set #2011 (used for model airplanes and radio-controlled cars)	2 set/pkg.

Item	Description	Quantity
Wheels	1" dia tailwheel for .60 size airplane, 3/32" axle dia; DU-BRO Co., #100TW	
Connector pin set	W.S. Deans Co. #3512 (used for model airplanes)	1 pkg.
Micrometer	A. Select a micrometer and matching gearhead from the catalog provided by The Little Depot B. Micrometer with built-in gearhead used in HO gauge Grandt Line Boxcab Co. Powering Kit, Walthers Co.	
Solar cells	.55 V/.3 A solar cells; Radio Shack Co.	3-6 cells
Ultra-fine connector wire	Radio Shack Co.	
Axle	3/32" brass rod; K&S Engineering Products	2
Spur gear rod and support rod	1/16" brass rod K&S Engineering Products	2
Photo matte board		
Solder		
Epoxy glue		
Super Glue		

CHAPTER
SEVEN

Solar Music Box

THE solar-powered music box (Figure 7-1) is an enjoyable introductory activity that explores the characteristics of photovoltaic cells. Its three cells are wired in series to provide enough output (1.65 volts) to power a miniature circuit and speaker that plays a medley of songs when light strikes the solar cells. A small 35 mm film container houses the circuit/speaker, with the three cells cemented to the outside of the unit.

The miniature circuit used in the project is an electronic circuit "recycled" from the inside of a greeting card or obtained from a micro-circuit supplier. You can find circuits that play Christmas and other assorted types of songs. (See the parts list.)

Once assembled, the music box is very sensitive and will start to play when exposed to even indirect sunlight. When placed on a window sill, it will really "hum" a tune when hit by full sunlight. The unit will also start to play if placed under a lamp with a 100- to 150-watt light bulb.

Construction Procedure

1. Carefully review Figures 7-1 through 7-4 (photos of components and the completed music box, a wiring diagram, and a three-view drawing of the project).

2. Cut three 1" lengths of fine wire. Next, cut two 3" pieces of wire. Strip the insulation off the ends of each wire.

3. Following the wiring diagram in Figure 7-2, carefully solder one of the 1" wires to the negative terminal on top of the first solar

Figure 7-1. This solar electric music box plays a medley of song when exposed to sunlight. The project is powered by three solar cells and uses a recycled microcircuit and a 35 mm film container.

cell. Next, solder the other end of the wire to the positive side of your second cell. Solder another piece of 1" wire to the negative terminal of this second cell, then solder the other end to the positive side of your third solar cell. This procedure will connect your cells in a series circuit, adding the voltage of the three cells together.

4. Solder a piece of the 3" wire to the positive side of your first solar cell. Next, solder the other 3" length of wire to the negative terminal of your third cell. These two wires serve as the positive and negative leads to your microcircuit.

5. Carefully push the three solar cells together by bending and curving their connecting wires (Figure 7-3). Set the array of solar

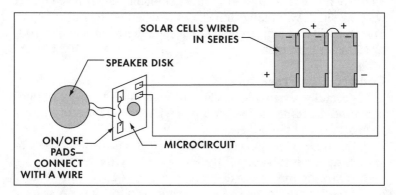

Figure 7-2. Schematic

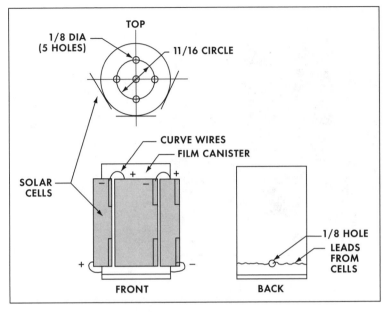

Figure 7-3. Three-view drawing

cells aside until you have modified your film canister housing and microcircuit.

6. Four holes must be drilled in the 35 mm film canister. The hole locations are not critical. Figure 7-3 illustrates the size and general location of the openings. The four holes on top, 1/8" in diameter, allow the sound to come out of the canister. The 1/8" hole in the back allows for connection of the wires from your solar array to the microcircuit.

7. The next phase of the project requires an electrical adjustment to your particular microcircuit. As illustrated in Figure 7-4, the musical circuits are based on three components: (1) a battery, (2) a microchip circuit, and (3) a disk speaker element.

8. Remove the battery system from your microcircuit. This usually involves either "popping off" the battery clip or cutting the lead wires from the circuit to the battery. Mark the positive and negative leads with a pen to make sure you can identify them.

9. Some microcircuits have an on/off system that must be permanently connected with a wire soldered between the circuit's pads (Figure 7-2).

10. Place a small amount of epoxy cement on the center of the speaker disk and bond it inside the top of the 35 mm film canister.

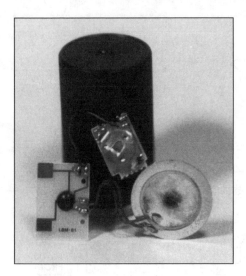

Figure 7-4. This photo illustrates the microcircuit, battery (center), speaker disk, and 35 mm film canister. The battery must be removed from the system—the solar cells will serve as the power source.

You will have to gently place the microcircuit inside the canister since it is connected to the speaker disk. You do not have to glue the microcircuit inside the container.

11. Once the speaker disk has dried, you can bond the three solar cells to the outside of the film canister (Figure 7-3). Carefully glue the center solar cell in place using a few drops of Super Glue. Next, glue the left cell in place, using care to push it close to the center cell. Finally, glue the right cell in position, pushing it close to the edge of the center cell.

12. Once the glue has dried, slide the positive and negative 3" wire leads from the solar cells through the hole in back of the film canister. Gently pull out the microcircuit and solder the solar cell leads to the positive and negative terminals on the circuit board.

13. Carefully push the microcircuit and any slack wire back inside the film container. Snap on the cap and the project is ready for testing.

14. To test the music box, place it in direct sunlight. You should hear it begin to play its tune. Place the unit on a table under a lamp fitted with a 100 watt to 150 watt light bulb, and it should play its medley. Have fun testing your project under different light conditions.

Alternative Activities

You can add interest by modifying the music box in the following ways:

● Decorate the music box to fit the type of music its microcircuit plays.

● Design your unit as an item intended to be displayed on a window sill or table (such as a Christmas ornament).

● Experiment with different ideas for housing the circuit and solar cells.

PARTS LIST		
Part	**Description**	**Quantity**
Solar cells	.55 V, .3 A type; Radio Shack	3
35mm film container	Use the disposable 35 mm film canister used to package film (Check your local photo store; some have extra containers.)	1
Musical microcircuit	Check musical birthday and holiday cards at Hallmark card stores. Recycle the microcircuit. Also, Klockit Co. has microcircuits used in projects that can be taken apart and recycled for solar projects.	1
Fine hookup wire	Use fine computer wire; Radio Shack	1
Solder	Radio Shack, use a thin solder for electrical work	10" length
Epoxy glue and Super Glue		

CHAPTER EIGHT

Solar Rechargeable Flashlight

THE field of photovoltaics has advanced with improvements in rechargeable batteries. In a variety of applications, solar cells are used to recharge ni-cad batteries that provide energy for products that operate during evening hours. The solar rechargeable flashlight project illustrates this process.

This activity actually comprises two separate projects that work together: a basic battery-powered light circuit and a solar cell module (Figure 8-1). The light project has four basic components: (1) a 1.25 V rechargeable ni-cad battery, (2) a miniature spst switch, (3) a small 1.5 V bulb, and (4) a miniature socket. The solar module consists of four .55 V, 300 mA solar cells connected in a series circuit and mounted in a plastic box.

The completed project serves as a pocket flashlight or, because of its compactness, you can easily keep it in your car's glove compartment. The light is charged by connecting the solar cell module to the light circuit using the miniature plug and socket, and placing the unit under a light source. With either bright sunlight or a 75- to 100-watt light bulb, charging will take three hours. Once charged, the pocket flashlight will operate *continuously* for five and a half hours.

Construction Procedure—Rechargeable Light Circuit
1. Carefully review the photograph of the light (Figure 8-1) and

the component layout/wiring diagram (Figure 8-2) before you begin construction.

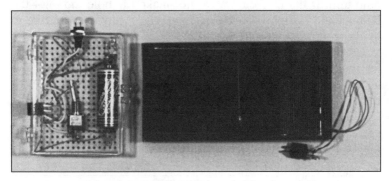

Figure 8-1. Completed light with solar charger.

2. A 1-5/8" x 2-1/8" x 5/8" plastic box serves as housing for the components. Drill a 1/8" diameter hole into the top/side section of the box (Figure 8-2).

3. Enlarge the hole into a rectangular 5/32" x 3/16" opening for the miniature socket. Use a small square file and small flat file for this operation.

4. Push the socket into the opening, with the face of the socket protruding a small distance from the box. Place a tiny drop of Super Glue around the edges of the opening to secure the socket in position.

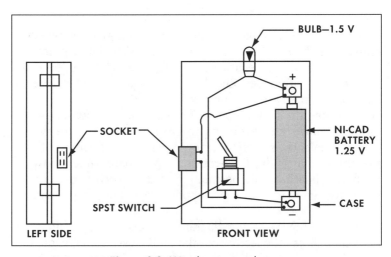

Figure 8-2. Wire/component layout

5. Drill a 1/8" hole in the top/center section of the box for the light bulb. Using a small round file, enlarge the opening for the bulb until it fits in position. Do not make the hole too big—the bulb should fit tightly in the opening. Use a tiny drop of glue to hold the unit in place.

6. Cut a piece of perfboard to 9/16" x 2-1/16". Cut a rectangular opening in the perfboard 7/16" x 1-1/16". This opening allows an N-size ni-cad battery to fit inside the small plastic box. Place the opening on the right side of the perfboard (Figure 8-2).

7. The terminal clips for the battery are made from two spring clips. Trim away the outside portion of the clips with a pair of diagonal cutters, and use the inside portion only (Figure 8-3).

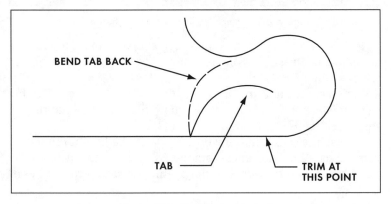

Figure 8-3. Spring clip terminal

8. Place the N-size battery in position, then glue the two clips onto the perfboard. Place one against the positive battery terminal and one against the negative terminal. These clips can be glued into position with Super Glue; however, make sure *not* to glue the battery onto the perfboard or case.

9. Turning the light on and off requires installation of an spst switch. Remove the two mounting nuts from the top of the switch. File the side of the switch to remove the remaining round collar so that the switch can lay flat against the perfboard. Glue the switch to the perfboard. While positioning of the switch is not critical, you may want to refer to the component layout drawing (Figure 8-2).

10. Following the wiring diagram, connect each of the four major components. The leads from the bulb must be long enough to allow for opening and closing of the box. Leads from the socket to

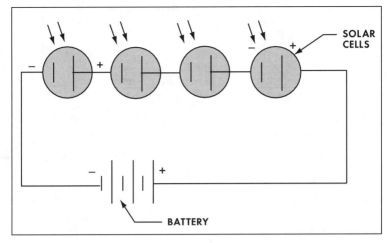

Figure 8-4. Schematic for solar cell charger.

the battery must also be long enough to allow opening and closing of the plastic box.

11 Carefully solder all of the wires in position.

12. Test the connections of your light by placing a standard alkaline N-size battery in the unit. Turn on the switch, and the unit should light. If the project does not operate, check all solder connections.

Note: This project can be used in a conventional manner by substituting a standard alkaline N-size battery for the rechargeable ni-cad battery.

Construction Procedure—Solar Cell Charger

1. Carefully review the photograph (Figure 8-1), the schematic drawing (Figure 8-4), and the case parts drawing (Figure 8-5) of the solar charger before beginning construction.

2. Following the schematic drawing, carefully solder four solar cells in a series circuit. Each solar cell produces .55 volt at 300 mA. Thus, total output of your charger will be 2.25 volts.

3. To connect the cells in a series circuit, solder a piece of fine-gauge wire (1/2" in length) from the negative terminal of the first cell to the positive terminal of the second cell. Next, solder a piece of wire from the negative terminal of the second cell to the positive terminal of the third cell. Finally, solder a wire from the negative terminal of the third cell to the positive terminal of the fourth cell.

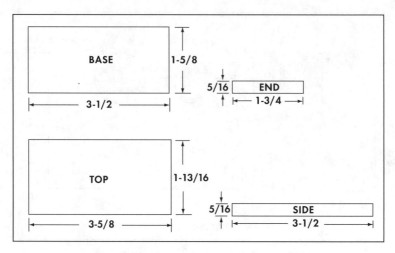

Figure 8-5. Case components

4. Solder a 6" length of wire to the negative terminal of the fourth cell. Next, solder a 6" length of wire to the positive terminal of the first cell. These wires will serve as your lead wires from the charger unit.

5. Cut a piece of matte board to a finished size of 1-5/8" x 3-1/2". Carefully push the cells together, so that the edges are almost touching. You will have to gently curve the wires that connect the cells. The solar cells should now fit on the matte board.

6. Use epoxy bond glue to attach the cells to the matte board. Handle the cells carefully—they will crack if you place too much pressure on them. (Note: Construction of a solar module was described and illustrated in the solar rover project in Chapter 6. Review Figure 6-7 of that project for additional information regarding connecting and mounting of solar cells.)

7. Set the solar module aside to allow the glue to dry.

8. Construct the case for the solar cells from 1/16"-thick black Plexiglas (Figure 8-5). Cut the base first, to a size of 1-5/8" x 3-1/2". Sand the edges of the base smooth.

9. Next, cut two side pieces of Plexiglas. Each side is 5/16" x 3-1/2". Sand the side pieces until the edges are smooth and straight.

10. Using a few drops of Super Glue, bond the side pieces to the edges of the base (Figure 8-6).

11. Cut two end pieces, each 5/16" x 1-3/4". Sand the end pieces until they are smooth and straight. Bond the two end pieces into position.

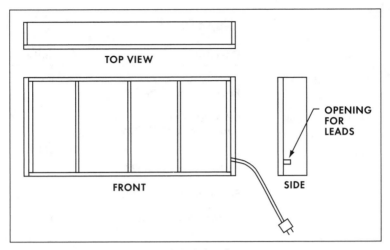

Figure 8-6. Solar charger

12. One end of the case will require an opening for the wire leads from the solar cells. File a 1/16" rectangular opening in the right end of the plastic case. The side of a small flat pattern file works well for creating this opening. The opening for the leads should be 1/8" from the edge of the case (Figure 8-6).

13. Carefully epoxy bond your solar cell panel into the plastic case. Gently run the wire leads along the bottom edge of the case and then out the opening at the case's end. Set the assembly aside to allow the glue to dry.

14. The top of the charger's case is made from a piece of 1/16" clear Plexiglas. Cut the plastic to a finished size of 1-13/16" x 3-5/8". Smooth the edges with a piece of #180 or #220 finishing paper.

15. Place a drop of epoxy glue at each corner of the case, then carefully place the clear plastic top in position. Allow the glue to dry.

16. Solder the positive and negative leads from the solar charger to the miniature plug that matches the socket in your light project. Make sure that the positive lead from the charger connects to the positive lead of the battery when you plug the charger into the light unit.

17. Set both units in bright sunlight for three hours. The units can also be charged near a 75- to 100-watt bulb for three hours. (Do not try to charge the unit under *overcast* skies.)

18. Do not leave the charger unit connected to the light project overnight or the battery will lose its charge.

PARTS LIST

Item	Description	Quantity
Silicon solar cells	.55 V, .3 A; Radio Shack #276-124	4
Miniature lamp	1.5 V, 25 mA; Radio Shack #272-1139	1
Switch	Micromini, spst; Radio Shack #275-624	1
Perfboard	Radio Shack #276-1395	1
Plastic box	1-3/4" × 2-1/4" × 11/16"	1
Spring clips		2
Connector pins	W. S. Deans plug and socket connector pins	1 set
ni-cad battery	1/25 V ni-cad; Radio Shack #23-121	1
Wire	Fine gauge wire	3'

CHAPTER NINE

Solar-Powered FM Transmitter

MANY remote regions in the world lack the electricity needed to operate a variety of products. To power transmission stations in these areas, engineers have developed photovoltaic panels that provide electricity for radio and TV transmitters. An enjoyable project that simulates this system is the small FM radio transmitter described here (Figure 9-1). The unit is constructed from very basic electronic components and assembled on a small printed-circuit (PC) board. The transmitter is powered by a compact solar panel that produces somewhere around 1.96 to 2 volts of electricity (Figure 9-2). (The solar module that powers the transmitter is constructed from the plans illustrated in the solar-powered light project described earlier. See Chapter 8 for construction details.)

The FM transmitter will broadcast a distance of approximately 50 feet to a standard FM radio. The unit's solar panel will power the transmitter when placed under either bright sunlight or a 75- to 100-watt light bulb.

Construction Procedure

1. Carefully review the photo of the transmitter (Figure 9-1), the component layout photograph (Figure 9-3), the component layout drawing (Figure 9-4), and the schematic (Figure 9-5).

2. You will need a small soldering pencil (2/40 watt), solder, a small pair of pliers, and a pair of tweezers. A pair of wire strippers

Figure 9-1. The transmitter will broadcast to a standard FM radio up to a distance of 50 feet. The project uses readily available components assembled on a PC board.

Figure 9-2. The FM transmitter is powered by this solar panel, which produces about 2 V of electricity.

Figure 9-3. Component layout view of the transmitter. Layout is not critical, but this photo, used with the component layout drawing, can help minimize construction problems.

will also aid construction operations.

3. The transmitter is assembled on a small Radio Shack PC board. Figures 9-3 and 9-4 show a suggested component layout. Using these drawings and the wiring diagram (Figure 9-5), solder capacitor *C-1* (100 pF) to the circuit board.

4. Next, solder transistor *T-1* (NPN-MPS 3904) in position. Note: *Be sure to hold each pin on the transistor with pliers while you solder it. The pliers act as a heat sink to prevent damage to the transistor. Make certain that the correct pins, emitter, base, and collector are placed in their proper position on the circuit board.*

5. Following the wiring dia-

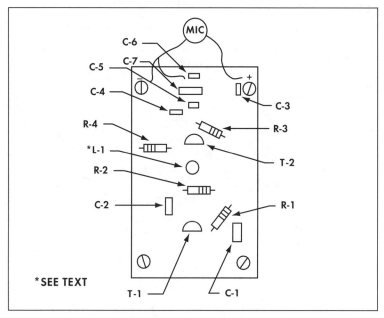

Figure 9-4. Component layout

gram, connect capacitor *C-1* to the collector pin of the transistor. Twist the wires together but do not solder yet.

6. Place resister *R-1* (150 ohms) in position. Connect one lead to the collector pin of the transistor. Do not solder at this time.

7. Place resistor *R-2* (18 K) in position. Twist one of its leads around the transistor's collector pin. Twist its other lead around the transistor's base.

8. Make sure that all of the connections are wrapped tightly. Holding the collector lead of the transistor with pliers (heat sink), solder *C-1*, *R-1*, and *R-2* to the collector lead.

9. Holding the base lead of the transistor with pliers, solder the other lead, *R-2*, to the base lead of the transistor.

10. Place capacitor *C-2* (100 pF) on the circuit board. Twist a piece of small-gauge wire around the lead of the capacitor that will connect to the transistor's base pin. Solder the lead to the capacitor. Run the wire to the transistor's base lead, wrap it tightly, then solder into place using the pliers.

11. Solder a 2-1/2" piece of small-gauge wire to the emitter lead of the transistor. Run the wire on top of the circuit board to any convenient hole at the other end of the board. Keep the wire on the left of the board. (You will solder the wire to components later.)

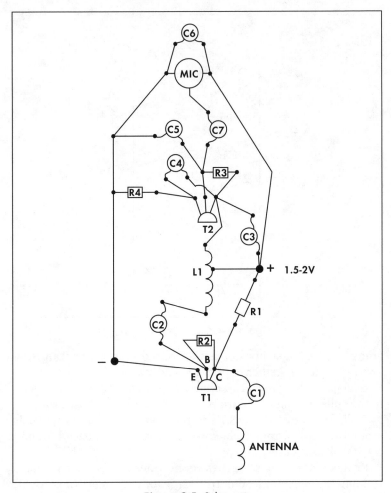

Figure 9-5. Schematic

12. Solder the other end of resistor *R-1* in position. Solder a fine-gauge piece of wire to this end of the resistor. Run the wire, about 2", to the other end of the circuit board. Keep the wire on the right side of the board.

13. Place capacitor *C-3* (4.7 pF) in position. Solder the wire from resistor *R-1* to capacitor *C-3*. Solder the wire to *C-3*'s lead that is toward the top of the board.

14. Put transistor *T-2* in position on the circuit board. Make sure that the collector, base, and emitter pins are in their correct position. Using your pliers as a heat sink, solder the transistor in place.

15. Solder the second lead of capacitor *C-3* (4.7 pF) to the collector pin of transistor *T-2*.

16. Set the partially completed transmitter aside. At this point, construct the coil, *L-1*, following the diagram in Figure 9-6. The coil is made from six turns of 22- or 24-gauge wire wrapped around a 1/8"-diameter form. The finished height of the coil should be approximately 1/4". You can use a small-diameter soda straw as the wrapping form for the coil. (As an alternative, a 1/8"-diameter screw will also serve as a form.)

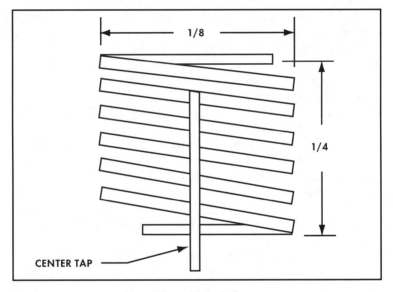

Figure 9-6. Coil

17. Coil *L-1* has a wire soldered to the fifth coil from its bottom. This is the *center tap* format for a coil. Note that the *L-1* coil is a very important component of the transmitter. In part, it determines the transmission frequency. If your project does not work when completed, check the coil construction. (You may have to construct a new coil.) Also, check to see if you have a good connection on the center tap wire to the coil.

18. Next, you can insert your completed coil in the circuit board. The second lead of capacitor *C-2* should now be soldered to the lead of the coil that is the "top" of the coil. (See the wiring diagram, Figure 9-5.)

19. Solder the center tap wire of the *L-1* coil to the junction of resistor *R-1* and capacitor *C-3*. (See Figure 9-5.)

20. The bottom lead of coil *L-1* should be soldered to the collector lead of transistor *T-2*. Remember to use a heat sink.

21. Place resistor *R-3* (18 K) in position. Solder one of the resistor's leads to transistor *T-2*'s collector pin. Solder the resistor's other lead to the transistor's base lead.

22. Insert capacitor *C-4* (10 pF) in position on the board. Solder one of its leads to transistor *T-2*'s collector pin. Then, solder its other lead to transistor *T-2*'s emitter pin.

23. Place resistor *R-4* (100 Ω) in position on the board. Solder one of its leads to transistor *T-2*'s emitter lead, then solder its other lead to the end of the wire that was run on top of the board and connects to transistor T-1's emitter pin.

24. Next, place the capacitor *C-5* (.0047 μF) on the circuit board. Then, place capacitor *C-7* (.047 μF) on the board. Solder one lead of capacitor *C-5* to the base pin of transistor *T-2* and solder its other lead to resistor *R-4*. Now, solder capacitor *C-7* to the base pin of *T-2* and solder its other lead to the "white" wire lead of the microphone element.

25. Solder the negative lead of the microphone element to the *C-5*, *R-4* connection. Solder the positive lead of the mike to the *C-3*, *R-1* connection. Check the schematic (Figure 9-5) for the illustration of these connections.

26. Solder capacitor *C-6* (.0047 μF) to the circuit. One lead connects to the mike's negative lead, and the other connects to the mike's positive lead.

27. Following the schematic, connect an 8" piece of wire to the negative input, then solder an 8" piece of wire to the positive input. The positive and negative leads from the transmitter should be connected to the positive and negative leads of a solar panel. (Again, the construction of a solar cell panel to power your transmitter is described in Chapter 8.)

28. Solder a 6" wire to capacitor *C-1* to serve as an antenna.

Operation and Troubleshooting Procedure

1. Place your solar panel in bright sunlight or under a 100-watt bulb, 6" to 8" from the bulb.

2. While you are tuning up and down the dial of your FM radio, speak into the mike element. At a blank location on your radio, toward the upper end of the FM dial, you should hear your voice.

3. If you do not hear your voice, consider the following:

● Was the *L-1* coil correctly made?

● Check all your connections. Are they soldered well?

● Check the schematic. Are all the components correctly wired together?

PARTS LIST		
Item	**Description**	**Quantity**
Microphone element	Replacement condenser mike element; Radio Shack #270-092	1
PC board	General purpose PC board; Radio Shack #276-149	1
Transistor T-1 and T-2	npn type, MPS 3904; Radio Shack #276-2016	2
Capacitor C-1 and C-2	Ceramic disk capacitor; 100 pF; Radio Shack #272-123	2
Capacitor C-3	Ceramic disk capacitor, 4.7 pF; Radio Shack #272-120	1
Capacitor C-4	Ceramic disk capacitor; 10 pF; Sprague Co., 10T Cc-Q10	1
Capacitor C-5 and C-6	Ceramic disk capacitor, .0047µF; Mouser Electronics #212-2142-472	2
Capacitor C-7	Ceramic disk type, .047 µF; Radio Shack #272-1068	1
Resistor R-1	150 Ω, 1/4 W; Radio Shack #271-1312	1
Resistor R-2, R-3	18 kΩ, 1/4 W	2
Resistor R-4	100 Ω, 1/4 W; Radio Shack #271-1311	1
Coil L-1	22- or 24-ga. enamel-coated wire (Farmor wire)	

APPENDIX A

Efficiency of Solar Cells

ABUNDANT energy exists for use in energizing photovoltaic cells. However, not all solar energy that strikes a solar cell is converted into electricity—solar cells are not 100 percent efficient. The efficiency of a photovoltaic cell is a measure of how well the cell can convert the sunlight that strikes it into electrical energy.

The following activity introduces the process of calculating the energy efficiency of a solar cell:

1. Create a simple circuit with a solar cell and motor (Figure A-1. See page 88.). For this example, we use a solar cell whose output is .55 V at 300 mA (Radio Shack #276-124). Connect a small motor (Pitsco #260) that draws a small amount of current to the cell, creating a simple series circuit.

2. Set up the circuit to obtain maximum power output (i.e., maximum volts and amps).

3. Under bright sunlight, your cell voltage should be .55 V at 300 mA (if you are using the Radio Shack #276-124 cell).

4. Calculate the cell's power as follows:

$$P = \text{voltage x current}$$
$$P = .55 \text{ V x } .3 \text{ A} = .165 \text{ W}$$

5. Solar insolation is the amount of sunlight that falls on your cell in your particular geographic region. For this example, use 255 Btu per sq. ft. per hour as the solar insulation and a solar cell that is .75" x 1.50". (Note that there are 144 sq. in. per l sq. ft., so

the cell area in square inches must be divided by 144 to obtain the Btus per square foot.)

6.   *Btus = solar insolation x solar cell area*
                                    144

    Btus = 255 x .75" x 1.50"
                          144

    Btus = 1.992

7. Given that 1 Btu = .2929 W, in our example,

    input power (in watts) = .2929 x 1.992
    input power = .583 W

8. Recall that power output of your cell in step 4 was = *.165 W*

9. Efficiency of cell = <u>*output power x 100*</u>
                           input power

    efficiency = <u>.165 W x 100</u>
                  .583 W
    efficiency = 28 percent

The example given here shows the *theoretical* efficiency of the solar cell. You can determine the actual value by placing the cell under a load and measuring its *voltage* and *current* output while it is driving a motor in bright sunlight. (See Figure A-1, page 88.) Using a voltmeter and an ammeter, you can measure the voltage and current and then insert them in the calculations above (at step 4). Note that you should connect the voltmeter in parallel to the cell. The ammeter must be in series *within* the circuit.

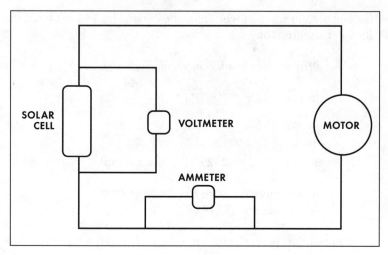

Figure A-1. Circuit

Types of Photovoltaic Systems

PHOTOVOLTAIC SYSTEMS

Stand-Alone Systems with dc Loads.	Stand-Alone Systems with ac Loads.	Grid-Connected Systems with ac Loads.
• Battery storage required	• Battery storage required	• Battery storage not required
• Used for small power needs, primarily lighting, communication, entertainment, and resistive loads.	• Used for all typical loads, including induction motors.	• Used for all typical loads, including induction motors.

APPENDIX C

Selected Bibliography

Adler, David. "Amorphous Semiconductor Devices." *Scientific American*, May 1977, pp. 36-48.

Chalmers, Bruce. "The Photovoltaic Generation of Electricity." *Scientific American*, October 1976, pp. 34-43.

Edelson, Edward. "Solar Cell Update." *Popular Science*, June 1992, pp. 95-99.

"EFG Photovoltaic Technology for Electric Utilities." Billerica, MA: Mobile Solar Energy Corporation, 1992.

Fahrenbruch, A. L., and Bube, R. H. *Fundamentals of Solar Cells*. Academic Press, New York, 1983.

Fan, John C. C. "Solar Cells: Plugging into the Sun." *Technology Review* 80, August-September 1978, pp. 2-19.

Fickett, Arnold, Gillings, Clark, and Lovins, Amory. "Efficient Use of Electricity." *Scientific American*, September 1990.

"High Efficient Solar Electric Modules" (technical brief). Camarillo, CA: Siemens Solar Industries, 1991.

"How a Solar Cell Works," Billerica, MA: Mobile Solar Energy Corporation, 1992.

Kern, E. D., and Pope, M. D. "Development and Evaluation of Solar Photovoltaic Systems: Final Report." MIT Lincoln Laboratory, May 1983. DOE/ET/20279-240.

Maycock, Paul, and Stirewalt, Edward. *Photovoltaics: Sunlight to Electricity in One Step*. Andover, MA: Brick House Publishing, 1981.

Peppriel, Barbara, ed. *Photocom Inc: Solar Electric Power Systems*

Design Guide. Scottsdale, AZ: Photocom Inc., 1991.

"Photovoltaics for Electric Utilities." Bellerica, MA: Mobile Solar Energy Corporation, 1989.

Rensberger, Boyce. "Consortium Seeks New Battery for Autos of the Future." *The Washington Post*, 19 June 1992.

Robertson, Edward. *The Solarex Guide To Solar Electricity*. Fredricksberg, MD: Solarex Corporation, 1979.

Russell, Miles. *Residential Photovolatic System Handbook*. Reading, MA: Sundance Publications, 1984.

Stahl, Dietrich. "Products with a Place in the Sun." *Siemens Review*, 57(1990), No. 6. (Published by Siemens Solar Industries, Camarillo, CA.)

Stepler, Richard. "Solar Electric Homes 1." *Popular Science*, September 1981.

Wardel, Charles. "Making a Solar Cell." *Fine Homes Builder, 72* (March 1992).

APPENDIX D

Solar Energy Terminology

Array—A group of photovoltaic modules, electrically wired together and mechanically installed in a structure or other common support system. The array functions as a single electricity-producing unit.

Alternating Current (ac)—Electric current that reverses its direction of flow 60 cycles, or 120 times, per second.

Amorphous Structure—A material in which there is no repeatable organization of atomic elements; a noncrystalline structure.

Ampere (A)—A measure of electric current. One ampere is created when an electrical force of one volt acts across a resistance of one ohm (Ω).

Barrier—In a photovoltaic cell, a thin region of static electric charge between the positive and negative layers. Also known as the cell *junction*. The barrier allows only high-energy electrons to pass through this zone, creating a current and voltage in the solar cell.

Boron—A semimetallic material that is used as a dopant to make positive, p-layer silicon in a solar cell.

British Thermal Unit (Btu)—The amount of heat needed to raise the temperature of one pound of water one degree Fahrenheit. One Btu equals 252 calories, 778 foot-pounds, 1,055 joules, and .293 watt-hours. It also equals .25 kilo calorie.

Concentrator Module—A photovoltaic module design that uses a lens to gather sunlight over a wide area and focus it on a smaller solar cell area.

Conversion Efficiency—The ratio of the power output of a photovoltaic cell to the incident power from the sun.

Current—Electricity, or the flow of electrons.

Direct Current (dc)—Current in which electrons are flowing in one direction only.

Dopant—An element added to a pure crystal of silicon to change its electrical characteristics. Boron and phosphorus are dopants added to silicon to create positive- and negative-layer (p-layer and n-layer) silicon.

Fill Factor—The ratio of the maximum power a solar cell can produce to the product of the open-circuit voltage and short-circuit current, as pertains to the current-voltage curve of a solar cell. Fill factor is a measure of the "squareness" of the current-voltage curve shape.

Flat-Plate Module—A grouping of solar cells in which the cells are exposed directly to normal incident sunlight.

Grid—A network of transmission lines, distribution lines, and transformers associated with a main power system.

Insolation—The solar energy received by a surface, per unit area, over a specified time period. Typical units are kWH/m per day.

Inverter—In a photovoltaic system, an inverter converts dc power from the photovoltaic array to the 60 Hz power that is compatible with the utility and house loads. Also known as a *power conditioner*.

Irradiance—The instantaneous incident solar power on a surface per unit area, kW/m.

I-V Curve—The curve that describes the fundamental electrical characteristic of a photovoltaic device, given as a plot of voltage versus current. The shape of the curve illustrates the solar cell's performance.

Kilowatt (kW) Peak—1,000 watts peak, with *peak* referring to noon on a sunny day.

Load—Any device in an electric circuit that uses power, such as an appliance.

Megawatt—One million watts.

Module—The smallest complete, protected assembly of electrically interconnected solar cells providing a single dc electrical output.

n-Silicon—Silicon with a dopant, such as phosphorus, added. (The phosphorus causes the silicon's structure to have more electrons than the number needed to complete the crystal structure.)

Open Circuit Voltage—The voltage measured across a solar cell when it is placed in sunlight; no current is flowing through the cell.

Peak Load—The maximum demand for electricity for a given period of time. Peak load is usually discussed in terms of the power used during a day.

Peak Watt—The quantity of power a photovoltaic cell will generate at noon on a clear day when the cell is tilted most favorably toward the sun.

Phosphorus—The element used as a dopant in silicon to produce negative silicon, or n-silicon.

Photovoltaic Array—See Array.

Photovoltaic Cell—A unit that converts sunlight directly into direct current electricity. Also called *solar cell*.

Photovoltaic Module—See Module.

Polycrystalline Silicon—A material that has a crystalline structure composed of randomly oriented segments of single-crystal lattice structure.

Power Conditioner—See Inverter.

PURPA (Public Utilities Regulatory Policies Act)—Federal legislation that mandates that utilities must allow small power producers (e.g., photovoltaic houses) to connect to the utilities' grids. Also requires that the utilities pay a fair price for surplus electricity created by the small producers.

p-Silicon—Silicon that contains the dopant boron. Boron does not provide sufficient electrons to complete the crystal structure.

Ribbon—Solar cell material grown in a long, narrow, and thin sheet. It can be either single-crystal or polycrystalline in structure.

Short-Circuit Current—The maximum current flowing through a solar cell; there is no load on the solar cell.

Single-Crystal—Describes a material that has a crystalline structure in which a repeatable or periodic molecular pattern exists in all three dimensions. A characteristic lattice is continuous throughout any size piece of the material.

Silicon—A semimetallic element that is commonly used as semiconductor material.

Solar Cell—See Photovoltaic Cell.

Substrate—The foundation or support layer of material in a photovoltaic cell.

Thin Film—A thin layer of a semiconductor material, such as silicon, that is drawn as a ribbon a few hundredths of an inch thick. Use of thin film offers a less-expensive way to manufacture solar cells.

Voltage—A measure of electrical potential in a circuit. The force of the flow of electrons in a circuit.

Watt—The electrical unit of power. One ampere of electrical current at one volt potential.

APPENDIX E

Directory of Suppliers

Most of the components specified in the projects in this book include the supplier's name and part number. The addresses of these suppliers are listed below. Many of these firms have a catalog available and will send a copy on request.

The parts list with each project is a suggested list of components needed. For convenience, you can substitute parts from another supplier.

Jacques Ebert Co.
Glen Cove, NY 11542

Kelvin Electronics Co.
10 Hub Drive
Melville, NY 11747
Phone: 516-756-1750
Recent catalog includes solar cells, motors, electronic components.

Klockit
P. O. Box 636
Lake Geneva, WI 53147
Microcircuits.

The Little Depot
1238-A South Beach Blvd.
Anaheim, CA
Phone: 714-828-5080
Recent catalog includes micromotors and gear heads.

Mouser Electronics
11433 Woodside Ave.
Santee, CA 92071-4795
Phone: 619-449-2222
Recent catalog includes electronic components and hardware.

Pitsco
1004 East Adams
P. O. Box 668
Pittsburg, KS 66762
Phone: 800-835-0686
Recent catalog includes solar cells, motors, gear sets, and other items.

Radio Shack
Parts available at all retail outlets.
For school orders:
Education Department
Radio Shack/Tandy
1400 One Tandy Center
Fort Worth, TX 76102
Recent catalog includes solar cells, resistors, capacitors, wire, other electronic components.

Walthers Co.
5601 W. Florist Ave.
Milwaukee, WI 53218
Micromotor with built-in gear head.

Local hobby shops can also provide supplies such as motors, K & S brass tubing (K & S Engineering Co.) and sheet materials, model servo gear sets, and so forth.

Index

References to figures, which help to define concepts contained in the text, are printed in boldface type. Definitions of many important terms can be found in the glossary (Appendix D).

Advanced Battery Consortium, 19
Aluminum-antimonide, 28
Amorphous materials, 41
Antireflection coating, **27**
Applied Solar Energy, 3
Array
  mobile, 8
  photovoltaic, 8, 11, 15, 29, 30, 34
  solar, 34
  solar cell, 5, 37
  solar electric, 7
Automated cell manufacturing, 42

Battery
  light activated, 22
  nickel-metal hybrid type, 19
Becquerel, Alexandre Edmond, 1
Bell Laboratories, 2
Boeing Corporation, 43
Boron, 25, 38
Boston Edison Co., 17

Cabin systems, 14
Cadmium-sulfide, 28
Cadmium-telluride, 28, 43
Calculators (solar powered), 10
Caltrans installation, 35
Capillary action, 40
Carlisle Home, **16**, 17
Charge regulator, 8
Chrysler Corporation, 19
Contact fingers, **27**
Conversion efficiencies, 5, 41
Copper-cuprous-oxide, 1
Corpuscles, 22
Crystalline silicon, 24. *See also* Silicon
Current, 30
Curacao, 9
Czocharalski process, 37-39

Desalination units, 13

Diamond abrasive saws, 38
Dopants, 38
Doping, 25

Edge-defined film-fed growth process (EFG), 39, 40
Efficiency
  as in energy conversion, 27
  practical, 27
Einstein, Albert, 1, 22
Electric Vehicle Amendment, 17
Electrolyte, 1
Electron-hole movement, 2, 24
Energy conversion devices (ECD), 42
Ethylene vinyl acetate poltant, 33, **34**

Federal Express Company, 20
Flaschglaff Solar Company, 44
Ford Festiva, **18**, 19
Ford Motor Co., 19
Fusion, 4

Gallium, 43
Gallium-antinomide, 43
Gallium-arsenide, 2, 28, 43
Gay, Dr. Charles, 4
Gell-type battery, 12
General Motors, 19, 20
Germanium alloy, 28, 42
Grain boundary, 39
Graphite die, 40

Hybrid vehicle, 17

Indium phosphide 2, 28
Induction coils, **40**
Infrared light, **23**
Insolation (sun intensity), 7
Irradiance, 28
Israel, 13
I-V curves, 29, **30**

Junction box, 33

Kvant, 43
Kyocera Company, 33

Laminator, **41**
Laser cutting system, 41
Lattice, 24, 25
Light-sensitive materials, 22
Low-iron glass, 34

Maximum power
  current, 29
  output, 29
  voltage, 29
Mazda Company, 44
Metallurgical-grade silicon, 44
Microgenerators, 11
Mini-panel, 47
Mobile Solar Corporation, 3, **26**, 39, 40
Module 10, 11, 17, 29, 30, 33
  fabrication, 38
  photovoltaic, 33, 34

Nanometers, 23
National Climatic Data Center, 7
National Energy Security Act 1991, 17
Nickel-metal hybrid battery, 19
N-layer, 26, 28, 30

Outer-shell electrons, 24
Ovonic Battery Company, 19
Ovshinsky, Stanford, 42, 43

Papua, New Guinea, 13, 14
Parallel circuit, 31
Parallel format, 31
Peak generating capacity, 3
Peak sun hour maps, 7
Phosphorous, 25, 38
  doped, 26
Photon(s), 1, 2, 22, 30, 31
Photovoltaic array. *See* Array
Photovoltaic cell, 1, 6, 10, 17, 18, 27, 29, 30, 39
  as energy for satellites, 2
  reflection on, 28
Photovoltaic effect, 1, 2, 23

Photovoltaic modules, 33, 34. *See also* Modules
Photovoltaic panel, 29
Photovoltaic panel pumps, 13
Photovoltaic process, 24, 37
Photovoltaic system
  remote, 9
  stand alone, 8, 13-15
  utility interactive, 16, 17
P-layer, 26
Polycrystalline silicon, 28, 37, 38
  sheet form, 39
  polysilicon, 39
Polymer, 33
Power inverter, 15, 16
Public Utilities Regulatory Policies Act, 15
Purification, 37

Radio Hoyer, 9
Radio Shack, 48
Rechargeable storage battery, 8
Recombination, 25
Recreational vehicle (RV), **10**, 11
Relativity, Einstein's theory of, 1
Resistance, 29.
Retrofit, 18
Roll-to-roll solar cells, **43**

Saudi Arabia, 13
Schuchuli Indian village, 12, 13
Selenium, 1
Semiconductors, 23, 24
Senate Energy Committee, 17
Series circuit, 31
Series/parallel circuit, 31-33
Sharp, Inc., 39
Shorts, electrical, 29, 39
Siemens Solar Industries, 3, 4, 14, 33-35
Silicon 24, 25, 28, 37, 38
  amorphous, 41, 42
  doped, 25
  metallurgical grade, 44
  molten, 49
  n-type, 26
  p-type, 25
  seed crystal, 40
  sheet, 41
  single crystal, 2, 38, 39

solar cell, 26-28, 30, 31
solar grade, 44
Society of Automotive
Engineers, 20
Solar array, 34. *See also* Array
Solar Car Corporation, 18
Solar cells, 37
roll-to-roll, **43**
Solar cell wafers, 38
Solar constant, 5
Solar Design Associates, 16
Solar electric car, 18
Solar electricity, field of, 1
Solar electric power
applications, 7
commercial products, 12
economics of, 2
Solarex Corporation, 3
Solar-grade silicon, 44
Solarpal, 12
Solar space panels, 12
Solar space radiation, 5
various forms, **6**
Southern California Edison, 44
Space program, U.S.A.
impact on science of solar
electricity, 3
satellites, 2, 3

Spheral solar electricity 44, 45
Sunrayce USA, **19**, 20

Tandem cells, **43**
Texas Instruments Company, 44
Thin-film solar cells, 44
Trickle charge, 44
Triple-cell structures, 42

Ultraviolet light, **23**
U.S. Coast Guard, 9, 10
U.S. Indian Health Service, 12
U.S. Postal Service, 20
U.S. Senate, 17
Utility grid, 16, 33
Utility-scale systems, 14-15

Vapor, 39
Visible spectrum, 22, 23
Voltage, 26, 28-32

Wafers, solar cell, 37, 38
Wavelengths, 4, 22
Western Electric, 2
World Solar Challenge, 20

Zero-emission vehicles, 17